Matrices

With

Applications

HUGH G. CAMPBELL

Virginia Polytechnic Institute

Prentice-Hall, Inc., Englewood Cliffs, New Jersey

TO
WILLIAM WEBB LOY
and
VIRGINIA THIGPEN LOY

©1968
by PRENTICE-HALL, Inc.,
Englewood Cliffs, New Jersey

Printed in the United States of America

ISBN: 0-13-565424-6

Library of Congress Catalog Card Number: 68-15227

10 9 8 7 6 5 4 3

PRENTICE-HALL INTERNATIONAL, INC., *London*
PRENTICE-HALL OF AUSTRALIA, PTY. LTD., *Sydney*
PRENTICE-HALL OF CANADA, LTD., *Toronto*
PRENTICE-HALL OF INDIA PRIVATE LIMITED, *New Delhi*
PRENTICE-HALL OF JAPAN, INC., *Tokyo*

Matrix algebra is a branch of mathematics that is becoming increasingly important to many disciplines. Current publications in agriculture, business, economics, political science, psychology, physics, chemistry, biology, and engineering reflect this fact.[1] The increasing emphasis on quantitative methods in all of these areas and the use of computers to perform matrix operations are two of many reasons for this development. Because of this obvious trend, it appears desirable to present this subject to students and their teachers at a much earlier stage than tradition has dictated.

The purpose of this book is to introduce matrix algebra to those who want an elementary, concisely written, survey of the subject. Although it is not intended to be a text on applied matrix algebra, it points the way to some elementary applications of matrices in the interest of motivating the reader. The book is suitable as a supplementary text for a variety of courses in agriculture, business, engineering, and the physical and social sciences. For those who want to introduce some matrix algebra into their present Freshman–Sophomore calculus sequence, this book will serve as a flexible companion text to supplement those calculus books with little or no matrix algebra. It is designed to serve, also, as a text for in-service training courses for secondary school teachers.

In its level and style this book differs from many introductory matrix texts. Also, because the book may be used in courses with few mathematics majors, the author has made a conscious effort to (1) motivate

---

[1] A quick survey of Part IV of the book *Applications of Undergraduate Mathematics in Engineering* by Ben Noble, The Macmillan Company, New York (1967), should convince the physical science or engineering student of the importance of matrix algebra. Ample material also exists for a similar book in the social sciences.

the reader with a subsection on applications for almost every section in the book; (2) immediately illustrate each new concept with an example; (3) as gently as possible, try to elevate the student's appreciation of mathematical structure; and (4) ease the pain of learning a new language by clearly defining each new term and listing the new vocabulary at the end of each chapter.

It is assumed that the reader has mastered elementary algebra and plane geometry. A knowledge of certain topics from elementary analytic geometry and calculus is assumed in a few parts of the book; if necessary, however, these topics can be skipped or explained sufficiently by the teacher.

The material has been arranged to permit maximum flexibility in course use. The Appendix contains a section on the summation notation and the proofs of some of the theorems; this material may be covered or omitted as the instructor desires. Also, all of the applications, except those that are starred, and Sections 2.5 and 6.2 may be omitted in the interest of brevity. And finally, for those who wish to get to Part III as soon as possible, Chapters 5 and 6 may be omitted; this procedure is not recommended, however.

Sincere appreciation is expressed to Professor Robert E. Spencer for his helpful suggestions in the preparation of the manuscript; to my wife, Allen, for typing the manuscript; and to Professor R. W. Brink for his careful editing and valuable suggestions.

H. G. C.

Blacksburg, Virginia

Contents

CONTENTS

Part II: SYSTEMS OF LINEAR EQUATIONS

Chapter 4

SYSTEMS AND RANK

Chapter 5

SYSTEMS WITH A UNIQUE SOLUTION

Chapter 6

SYSTEMS WITH MORE THAN ONE SOLUTION

Part III: MATRIX TRANSFORMATIONS

Chapter 7

LINEAR TRANSFORMATIONS

CONTENTS

Part: I

Matrix Operations

Basic Terminology

1.1 AN ORIENTATION

Concurrently with the development of civilization, men have found ways to quantify certain events, situations, or arrangements of matter of the real world. Such a mathematical representation is often called a *mathematical model*. One of the earliest examples of the construction of a usable mathematical model was the invention of the set of natural numbers $1, 2, 3, \ldots$ enabling people to count objects. In the seventeenth century Sir Isaac Newton (English, 1642–1727) and Baron Gottfried Wilhelm von Leibniz (German, 1646–1716) independently developed a rather sophisticated type of mathematics, known as the calculus, in order to describe certain observations about the physical and geometrical world. In other words, they constructed mathematical models of physical phenomena. This procedure enables a scientist to study the properties of a given physical or social phenomenon by studying an analogous mathematical model. From this study a better theoretical understanding of the working of the given system is gained and perhaps through this understanding new theories can be justified. An illustration of these ideas is given in the following example.

Example 1. Many years ago a mathematical model was developed in which there were two elements, 0 and 1, and three operations ($\cup$), ($\cap$), and ($'$). These operations were defined as follows:

$$0 \cup 0 = 0, \qquad 0 \cap 0 = 0, \qquad 0' = 1,$$
$$1 \cup 1 = 1, \qquad 1 \cap 1 = 1, \qquad 1' = 0.$$
$$1 \cup 0 = 0 \cup 1 = 1, \qquad 1 \cap 0 = 0 \cap 1 = 0,$$

From these basic definitions or assumptions many properties of this model can be proved. For example, if x represents either 0 or 1, and if y represents either 0 or 1, then

$$x \cup (x \cap y) = x,$$

and

$$(x \cap y)' = x' \cup y'.$$

This can be seen by comparing column (1) with column (5) and column (8) with column (9) of the accompanying table that lists all possible combinations.

(1)	(2)	(3)	(4)	(5)	(6)	(7)	(8)	(9)
x	y	$x \cup y$	$x \cap y$	$x \cup (x \cap y)$	x'	y'	$(x \cap y)'$	$x' \cup y'$
0	0	0	0	0	1	1	1	1
0	1	1	0	0	1	0	1	1
1	0	1	0	1	0	1	1	1
1	1	1	1	1	0	0	0	0

This model can be used to represent the physical arrangement of electrical switches. Let 0 represent an open switch, 1 represent a closed switch, $\cup$ indicate that the two switches are connected in parallel, $\cap$ indicate that the two switches are connected in series, and (') change the state of the switch from open (closed) to closed (open). This application of the model is extremely useful in electrical engineering, particularly in computer designs. This model can also be used in another way by letting 0 represent a false statement, 1 represent a true statement, $\cup$ represent the connective "or," $\cap$ represent the connective "and," and (') represent the negation of a statement.

We shall find that matrices may be used to construct many useful mathematical models. In particular, the purpose of this book is to introduce matrices sufficiently for them to be used in studying both general systems of linear equations and simple linear transformations. A further purpose is to begin building a vocabulary to enable the reader to read and understand literature which uses the language of matrix algebra. Although this is not a text in applied matrix algebra, many of the diverse applications are pointed out along the way.

Actually the concept of a matrix is not a new one; it was first introduced by the mathematician Arthur Cayley (English, 1821–1895) in 1858. Recently, the technological explosion and the high-speed computer have created a new interest in matrices. For example, certain problems in decision-making which have hitherto been very difficult for businessmen to solve are now quickly resolved by computers after a translation of the problem has been made into the language of matrix algebra. Thus, the application of matrices is becoming increasingly important to scholars in all branches of the biological, social, natural and engineering sciences. There is a great need for trained personnel who can bring the power of matrix methods to bear on the problems of these disciplines.

EXERCISES

1. (a) Referring to Example 1 of this section, prove that $y \cap (x \cup y) = y$.

(b) Express this relationship as an arrangement of switches (draw a diagram).

(c) Express this relationship as a sentence containing certain statements and connectives.

2. Simplify the following expression using the definitions of Example 1 of this section:

$$(y \cup x)' \cap (y \cup x').$$

1.2 BASIC DEFINITIONS

In constructing a new mathematical structure it is essential to begin with the definitions of the basic vocabulary.

Definition 1.1. *A matrix A is a rectangular array of entries (or elements) and may be denoted by*

$$A = \begin{bmatrix} a_{11} & a_{12} & \cdots & a_{1n} \\ a_{21} & a_{22} & \cdots & a_{2n} \\ \cdot & \cdot & & \cdot \\ \cdot & & \cdot & \cdot \\ \cdot & \cdot & & \cdot \\ a_{m1} & a_{m2} & \cdots & a_{mn} \end{bmatrix}$$

We shall assume that the entries of a matrix belong to the field of complex numbers unless otherwise specified.[1] We shall refer to such entries as scalars.

Example 1.

$$\begin{bmatrix} 2 & 3 \\ 4 & 0 \end{bmatrix}, \quad [\tfrac{1}{2} \quad 2 \quad 4], \quad \begin{bmatrix} \sqrt{3} & i & 0 \\ 1 & \tfrac{3}{2} & 2 \end{bmatrix}$$

are examples of matrices.

---

[1] Actually the entries may be the elements of any specified algebraic system and a corresponding theory of matrices can be developed.

If all of the entries of a matrix are zero, the matrix is called a *zero* or *null matrix* and is denoted by **0**. Bold print is used to distinguish a zero matrix from the zero scalar. If the entries of the matrix are real numbers, the matrix is called a *real matrix*. The matrix A, as given in the definition, consists of m rows and n columns. The subscript i of the entry a_{ij} designates the row in which the entry appears and the subscript j designates the column in which the entry appears; the double subscript ij is called the *address* of the entry. A matrix in which there are m rows and n columns of entries is said to be of *order "m by n"*; the number of rows is always stated first. A matrix is said to be a *square matrix of order n* if it has precisely n rows and n columns of entries. The *main diagonal* of a square matrix consists of the entries for which $i = j$. Frequently an m by n matrix A is expressed as

$$A = [a_{ij}]_{(m,n)};$$

this is referred to as the *abbreviated notation* where i varies from 1 to m and j varies from 1 to n.

Example 2.

The matrix $A = \begin{bmatrix} 4 & 2 & \sqrt{3} \\ 0 & 5 & 1 \\ 7 & \frac{3}{2} & 8 \end{bmatrix}$ is a square, real matrix of order 3, and the

entries 4, 5, 8 constitute the main diagonal. The entry a_{31} is 7 (third row, first column).

Definition 1.2. *If some rows or columns (or both) of A are deleted, the remaining array is called a submatrix of A. Also, it is customary to treat A as a submatrix of itself.*

Example 3. For

$$A = \begin{bmatrix} 3 & 2 & 1 \\ 4 & 6 & 9 \end{bmatrix},$$

we are able to obtain the following submatrices:

$$\begin{bmatrix} 3 & 2 & 1 \\ 4 & 6 & 9 \end{bmatrix}, \begin{bmatrix} 3 & 2 \\ 4 & 6 \end{bmatrix}, \begin{bmatrix} 3 & 1 \\ 4 & 9 \end{bmatrix}, \begin{bmatrix} 2 & 1 \\ 6 & 9 \end{bmatrix},$$

$$\begin{bmatrix} 3 \\ 4 \end{bmatrix}, \begin{bmatrix} 2 \\ 6 \end{bmatrix}, \begin{bmatrix} 1 \\ 9 \end{bmatrix}, [3 \ 2 \ 1], [4 \ 6 \ 9],$$

$$[3], [2], [1], [4], [6], [9], [3 \ 2], [3 \ 1], [2 \ 1], [4 \ 6], [4 \ 9], [6 \ 9].$$

This set of submatrices contains all of the submatrices for the given matrix A.

Sometimes it is convenient to divide a matrix into submatrices by means of broken lines. Thus we can write

$$A = \begin{bmatrix} A_{11} & \vdots & A_{12} \\ ---- & \vdots & ---- \\ A_{21} & \vdots & A_{22} \end{bmatrix} = \begin{bmatrix} a_{11} \; a_{12} & \vdots & a_{13} \\ a_{21} \; a_{22} & \vdots & a_{23} \\ -------- & \vdots & --- \\ a_{31} \; a_{32} & \vdots & a_{33} \end{bmatrix}.$$

When this has been done we say that A has been **partitioned.**

Definition 1.3. *Two matrices are said to be **equal** if they are of the same order and if all of their corresponding entries are equal; that is, $[a_{ij}]_{(m,n)} = [b_{ij}]_{(m,n)}$ if $a_{ij} = b_{ij}$ for all $i = 1, 2, \cdots, m$ and $j = 1, 2, \cdots, n$.*

Definition 1.4. *A real matrix A is said to be "**greater than**" ($>$) a real matrix B of the same order when each entry of A is "greater than" the corresponding entry of B. ($<$ or $\geq$ or $\leq$ can be substituted for $>$ with corresponding changes in meaning.)*

Example 4.   For matrices

$$A = \begin{bmatrix} 3 & 2 \\ 1 & 1 \end{bmatrix}, \quad B = \begin{bmatrix} 3 & 2 & 0 \\ 1 & 1 & 0 \end{bmatrix}, \quad C = \begin{bmatrix} 3 & 2 \\ 1 & 0 \end{bmatrix},$$

$$D = \begin{bmatrix} 1 & 1 \\ 0 & 0 \end{bmatrix}, \quad E = \begin{bmatrix} 3 & 3 \\ 0 & 1 \end{bmatrix}, \quad F = \begin{bmatrix} 3 & 2 \\ \frac{2}{2} & 1 \end{bmatrix},$$

we have the following relations $A \neq B$, $A \geq C$, $A > D$, $A \ngtr E$, $A \nless E$, $A \neq E$, and $A = F$.

APPLICATIONS

The following examples illustrate a few of the many ways in which a matrix is useful.

Example 5.   Consider the arrangement of four points (which might represent persons, or nations, etc.) shown in Figure 1.1. This diagram can be translated into the language of matrix mathematics by

	#1	#2	#3	#4
#1	0	1	0	1
#2	1	0	1	1
#3	0	1	0	0
#4	1	1	0	0

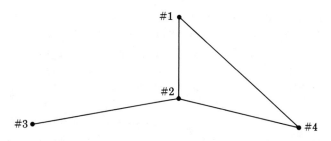

Fig. 1.1

where $a_{ij} = 1$ if the ith point is connected to the jth point and $a_{ij} = 0$ if these points are not connected. We assume that no point is joined to itself. This representation fully describes whether or not there is a line segment connecting the points (or whether there is communication between persons or nations, etc.).

For large communication systems, diagrams like Figure 1.1 are very unwieldy, but by performing various operations on associated matrices one may reduce the larger system to a simpler system or make some other scientific analysis. An interesting variation of this application may be found in Exercise 9 of this section.

Example 6. Suppose that there exists a certain defined relation between persons, nations, numbers, or biological characteristics. Let us call this relation "dominance," and in Figure 1.2 suppose that an arrow from point i to point j denotes the dominance of i over j. This dominance can be defined by the following matrix

$$
\begin{array}{c c}
 & \begin{array}{cccc} \#1 & \#2 & \#3 & \#4 \end{array} \\
\begin{array}{c} \#1 \\ \#2 \\ \#3 \\ \#4 \end{array} &
\begin{bmatrix} 0 & 1 & 0 & 0 \\ 0 & 0 & 1 & 1 \\ 0 & 0 & 0 & 0 \\ 1 & 0 & 0 & 0 \end{bmatrix}
\end{array} ,
$$

where $a_{ij} = 1$ if i dominates j and $a_{ij} = 0$ if i does not dominate j. Obviously, no point dominates itself.

A variation of this problem occurs when each point either dominates or is dominated by each of the other points. A very interesting use of this model was made in a study of influence among the justices of the Michigan Supreme Court.[2] The use of the concept of dominance may pertain to biological

---

[2] S. Ulmer, "Leadership in the Michigan Supreme Court" in *Judicial Decision Making*, G. Shubert, ed. (New York, The Free Press of Glencoe, 1963).

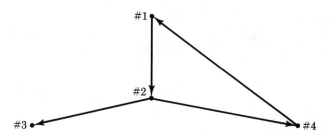

Fig. 1.2

characteristics, sociological behavior, and other areas as well as to political influence.

★[3] **Example 7.** One of the most valuable uses of matrices is in connection with systems of linear equations. For the system

$$\begin{cases} x + y + z = 4, \\ 2x - y + 3z = 6, \end{cases}$$

the matrix of coefficients is $\begin{bmatrix} 1 & 1 & 1 \\ 2 & -1 & 3 \end{bmatrix}$, and if this matrix is augmented

by the members of the right-hand side of the system, we have the matrix

$\begin{bmatrix} 1 & 1 & 1 & \vdots & 4 \\ 2 & -1 & 3 & \vdots & 6 \end{bmatrix}$. These are known, respectively, as the *coefficient*

matrix and the *augmented matrix* of the linear system.

Example 8. Suppose that a certain company owns four factories which produce two products. The amount of product i produced by factory j can be represented by the entry a_{ij} in a 2 by 4 matrix such as

FACTORIES

A B C D

PRODUCT #1 $\begin{bmatrix} 4 & 7 & 9 & 8 \\ 7 & 6 & 0 & 2 \end{bmatrix}$.
PRODUCT #2

---

[3] The starred examples should not be omitted as they form an essential part of this text.

Example 9. A standard problem in statistics is to study the effect that one variable has upon another in a given situation. The following matrix was taken recently from a statistical study made by a University.[4]

	HIGH SCHOOL RANK	SAT VERBAL	SAT MATH	MATH ACHIEVEMENT
HIGH SCHOOL RANK	1.00	0.28	0.19	0.22
SAT VERBAL	0.28	1.00	0.36	0.37
SAT MATH	0.19	0.36	1.00	0.73
MATH ACHIEVEMENT	0.22	0.37	0.73	1.00

Here a_{ij} represents the coefficient of correlation between the ith variable and the jth variable. A number near 1 shows a high degree of correlation, and a number near 0 shows a low degree of correlation.

EXERCISES

1. State the order of each of the following matrices

$$[2 \quad 1 \quad 3], \quad \begin{bmatrix} 1 & 2 & 4 & 0 \\ 4 & 0 & 3 & 0 \end{bmatrix}, \quad \begin{bmatrix} 2 \\ 3 \end{bmatrix},$$

and state the address of the entry "3" in each matrix.

2. List the entries on the main diagonal of $[a_{ij}]_{(4,4)}$.

3. Partition $\begin{bmatrix} i & 1 & 0 \\ 2 & 2 & \sqrt{3} \end{bmatrix}$ in such a way that a 2 by 2 real submatrix results.

4. Does the array of dates of a month on a calendar form a matrix? Why?

5. Write the 3 by 5 null (or zero) matrix.

6. List all of the 2 by 3 submatrices of

$$\begin{bmatrix} 3 & 4 & 2 & 1 \\ 0 & 2 & 1 & 4 \end{bmatrix}.$$

7. If possible, find all values for each unknown that make each of the following a true relationship.

$$(a) \begin{bmatrix} x & 2 \\ 3 & 4 \end{bmatrix} = \begin{bmatrix} 0 & \sqrt{4} \\ 3 & 4 \end{bmatrix}; \quad (b) \begin{bmatrix} 1 & \frac{4}{2} & t \\ 3 & 0 & 0 \end{bmatrix} = \begin{bmatrix} 1 & 2 & t \\ 3 & 0 & 0 \end{bmatrix};$$

[4] Virginia Polytechnic Institute Counseling Center, 1966.

(c) $\begin{bmatrix} 4 & 0 \\ y & 0 \end{bmatrix} = \begin{bmatrix} 4 \\ 4 \end{bmatrix};$ (d) $\begin{bmatrix} -1 & -2 \\ 4 & 3 \end{bmatrix} \leq \begin{bmatrix} 0 & -1 \\ x & 3 \end{bmatrix};$

(e) $\begin{bmatrix} 3i & 4 \\ 3 & x \end{bmatrix} > \begin{bmatrix} i & 2 \\ 1 & 2 \end{bmatrix};$ (f) $\begin{bmatrix} 2 & 0 \\ 4 & -3 \end{bmatrix} < \begin{bmatrix} x & 2 \\ 4 & -2 \end{bmatrix}.$

8. Make up an example of a 2 by 2 real matrix that is not greater than, less than, nor equal to $A = \begin{bmatrix} 0 & -2 \\ -3 & 6 \end{bmatrix}$.

9. Suppose that in Example 5 of this section we think of the lines connecting the points as communication lines and suppose that station #2 can speak to stations #1 and #4 but they cannot reciprocate. If we let arrows represent this one-way communication the diagram in Example 5 is modified as shown in Figure 1.3. Express this communication system in matrix form.

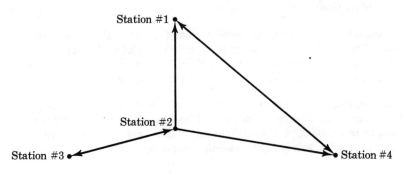

Fig. 1.3

10. One variation of the dominance concept illustrated in Example 6 is that *every* object either dominates or is dominated by *every* other object. Suppose we have this situation among four objects. Let the matrix representation be

	#1	#2	#3	#4
#1	0	0	1	1
#2	1	0	1	0
#3	0	0	0	0
#4	0	1	1	0

where $a_{ij} = 1$ if i dominates j and $a_{ij} = 0$ if i does not dominate j. Draw a diagram representing this dominance.

11. The augmented matrix of a certain system of linear equations is

$$\left[\begin{array}{cc:c} 1 & 3 & 1 \\ 4 & 2 & 5 \\ 2 & 6 & 4 \end{array}\right]$$. Write the system.

1.3 VECTORS

Matrices of order 1 by n are called **row matrices,** and matrices of order n by 1 are called **column matrices.** Note that row or column matrices with scalar entries are simply ordered sets of scalars. Here we use the word "scalar" to mean any complex number.

Definition 1.5. *A **vector** α of order n (or dimension n) is an ordered set of n scalars $(a_1, a_2, a_3, \cdots, a_n)$. In this book vectors are denoted by Greek letters $\alpha, \beta, \gamma, \cdots$.*

Example 1. The coordinates of a point (x, y, z) in three-space are an ordered set of scalars and can be considered as a vector. In fact, it is customary to represent vectors geometrically in three-space with an arrow from the origin to the point $P(x, y, z)$ as illustrated in Figure 1.4.

Example 2. The alphabetical price list of n items of a certain store constitutes an ordered set of elements, and hence is a vector. If the price list exceeds three items, a geometric representation similar to Example 1 is unattainable, but this is not necessarily a handicap.

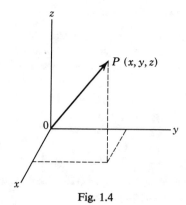

Fig. 1.4

In the definition of a vector the specific elements a_i of the ordered set are called **components** of α; vectors with n components are called **n-dimensional vectors**. A vector whose components are all zero is called a **zero** or **null vector**. If the components are all real numbers, then the vector is said to be a **real vector**. Often it is convenient to represent quantities from various applied fields as ordered sets of components, and then these quantities as vectors can be manipulated as single entities. Of course, in order to perform these manipulations certain definitions are necessary.

Definition 1.6. *Two vectors of the same dimension are* **equal** *if their corresponding components are equal.*

Definition 1.7. *A real vector α is said to be* **"greater than"** *($>$) a real vector β of the same dimension if each of the components of α is "greater than" the corresponding component of β. ($\geq$ or $\leq$ or $<$ can be substituted for $>$ with corresponding changes in meaning.)*

Notice that the preceding definitions concerning vectors are consistent with the definitions for matrices of the previous section and with the assertion that row or column matrices are vectors.

Definition 1.8. *Two vectors of the same dimension are* **added** *by adding corresponding components.*

Definition 1.9. *A vector is* **multiplied by a scalar** *by multiplying each component of the vector by the specified scalar.*

Definitions 1.8 and 1.9 are used to define subtraction of two vectors as

$$\alpha - \beta = \alpha + (-1)\beta.$$

Definition 1.10. *Let $\alpha = (a_1, a_2, \ldots, a_n)$ and $\beta = (b_1, b_2, \ldots, b_n)$ be two* **real** *vectors. The* **dot product** *or* **inner product** *or* **scalar product** *of α and β, denoted by $\alpha \cdot \beta$, is*

$$\alpha \cdot \beta = a_1 b_1 + a_2 b_2 + \cdots + a_n b_n.$$

[*Note: The dot should always be used between α and β in this particular type of multiplication because there exist other types of vector products.*]

Example 3. Let $\alpha = (4, 1, 1, 6)$ and $\beta = (3, 1, 0, -2)$. Then $\alpha \geq \beta$, but $\alpha \not> \beta$, $\alpha \neq \beta$.

$$\alpha + \beta = (4 + 3, \quad 1 + 1, \quad 1 + 0, \quad 6 - 2) = (7, 2, 1, 4),$$
$$\alpha \cdot \beta = 4 \cdot 3 + 1 \cdot 1 + 1 \cdot 0 + 6 \cdot (-2) = 1.$$
$$\alpha - \beta = \alpha + (-1)\beta = (4, 1, 1, 6) + (-3, -1, 0, 2) = (1, 0, 1, 8)$$

Note that the result of $\alpha \cdot \beta$ is a scalar. This is the reason for calling this operation the scalar product. Another type of vector multiplication yields a vector, and still another yields a matrix.

Theorem 1.1. *If α and β are two nonzero vectors in the x_1x_2-plane, then $\alpha \cdot \beta = |\alpha||\beta| \cos \theta$, where θ is the angle between α and β.*

Proof. See Theorem A.1, page 154.

Example 5 illustrates an application of this theorem.

APPLICATIONS

Since their invention in mathematics in the middle of the nineteenth century, vectors have become useful in many other disciplines. Herman Günther Grassman (German, 1809–1877) and Sir William Rowan Hamilton (Irish, 1805–1865) did many of the first studies of vectors, and their work provoked widespread interest in later years. The mathematical physicist Josiah Willard Gibbs (American, 1839–1903) did much work in applying vector methods to physics. In the present century it has become apparent that vectors may be used to provide useful mathematical models for problems of the social sciences as well as those of the natural sciences. Problem-solving methods using vectors are illustrated in the following examples.

★ **Example 4.** The quantification of certain physical notions such as velocity, acceleration, and force require well-defined mathematical entities that can be used to state both magnitudes and directions; vectors are ideally suited for this task. As mentioned previously a real vector $\alpha = (a_1, a_2, a_3)$ can be represented geometrically by an arrow from the origin to the point $P(a_1, a_2, a_3)$; the length of the arrow represents the magnitude of the vector. By the theorem of Pythagoras (Greek, died c 497 B.C.), the length of the arrow or the *magnitude* of the vector α, denoted by $|\alpha|$, is found to be $|\alpha| = \sqrt{a_1^2 + a_2^2 + a_3^2}$. This concept is generalized to define the magnitude of an n-dimensional real vector, $\beta = (b_1, \cdots b_n)$. $|\beta| = \sqrt{b_1^2 + \cdots + b_n^2}$. A vector that has a magnitude equal to 1 is called a *unit vector*.

Physical experimentation has demonstrated that for a particle resting on a plane, a force of $4\sqrt{2}$ units exerted on this particle in a northeasterly direction can be expressed as $\alpha = (4, 4, 0)$ (see Figure 1.5), a force of 3 units exerted in a westerly direction can be expressed as $\beta = (-3, 0, 0)$, and the result of exerting both these forces coincidently on the given particle can be represented mathematically by the sum or addition of the corresponding vectors. Thus, the resulting force is

$$\alpha + \beta = (4, 4, 0) + (-3, 0, 0) = (1, 4, 0);$$

the magnitude is $\sqrt{1^2 + 4^2 + 0^2} = \sqrt{17}$, and the direction is shown in Figure 1.5.

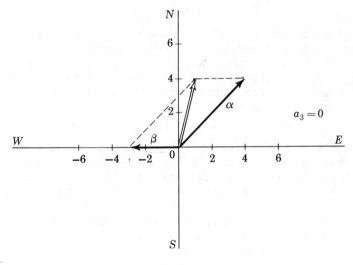

Fig. 1.5

Example 5. In the study of elementary physics we learn that as a particle moves through a certain distance, the work done on it by a constant force acting on it in the direction of movement is given by the equation

(work) = (force) (distance),

where the unit of work (such as foot-pound, dyne-centimeter, etc.) is expressed in terms of the units of force and distance. Suppose we want to find the work done (in a plane) in moving a particle along the path designated by $\alpha = (3, 1)$ by a force $\beta = (1, 2)$. (See Figure 1.6.) Since the total magnitude

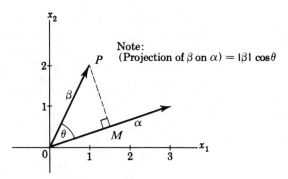

Fig. 1.6

of the force represented by the vector β, does not act directly in the direction of the particle's movement, we are interested only in the component of the force β that acts in the direction of α. From Figure 1.6 we see that this component is $|OM| = |\beta| \cos \theta$. The distance moved is $|\alpha|$; therefore, the total work done is $|\alpha| \, |\beta| \cos \theta$, which according to Theorem 1.1 is simply equal to $\alpha \cdot \beta$. Thus, we have found that

$$(\text{work}) = \alpha \cdot \beta = (3, 1) \cdot (1, 2) = 3 + 2 = 5.$$

Since the magnitudes of α and β are measured in different kinds of units, say $|\alpha|$ in feet (centimeters) and $|\beta|$ in pounds (dynes), there is actually no connection between the lengths of the two lines OP and OM. There is no reason why α and β need be drawn with coordinate axes of the same scale, except that in order to find θ we need to have their directions relative to each other. Let us, then, repeat this Example with the substitution $\beta' = (7, 14)$ with β' having the same direction as $\beta = (1, 2)$. The angle θ is therefore the same as before, and the new solution is

$$(\text{work}) = \alpha \cdot \beta' = (3, 1) \cdot (7, 14) = 21 + 14 = 35.$$

As we have expected, the work is 7 times as great as it was before, since all conditions remain unchanged except that the magnitude of the force was multiplied by 7.

Example 6. We live in a three-dimensional world; therefore many applied problems that are dealt with employ only three-dimensional vectors. However, n-dimensional vectors are used more often now than in the past; computers have played a big part in this development. Suppose an investment company decides to sell four different types of stocks. In one transaction 200 shares of stock A, 300 shares of stock B, 100 shares of stock C, and 200 shares of stock D are sold. The selling prices per share were $20, $30, $50, and $10, respectively. Let the total quantity of stocks sold be represented by the vector $\alpha = (200, 300, 100, 200)$, and let the selling prices be denoted by the vector $\beta = (20, 30, 50, 10)$. The total receipt from the stock sale is then

$$\alpha \cdot \beta = (200, 300, 100, 200) \cdot (20, 30, 50, 10)$$
$$= 20,000 \text{ dollars.}$$

Vector methods provide a convenient way to arrange large bodies of data and are becoming especially useful for problems of the above type where the dimension may be large. A computer can easily be programmed to find the dot product.

Example 7. Consider a precinct in which it is believed that of those who voted for a Democrat in the last election, 75% will vote for a Democrat in the next election, 20% will vote for a Republican, and 5% will vote for a third party candidate. The vector $(0.75, 0.20, 0.05)$ represents this prediction and is an example of a *probability vector*. A *probability vector* has non-negative components whose sum is 1.

Example 8. Suppose we construct a simple model economy in which there are three industries — the crude oil industry, the refining industry that produces gasoline, and the utility industry which supplies electricity. Then suppose that there are six types of consumers — the general public, the government, the export firms, and the three industries. Both the industries and the consumers exercise certain consumptive demands on each of the industries. For instance, suppose that the crude oil industry needs to use 4 units of gasoline and 2 units of electricity. The demand vector, then for the crude oil industry, is expressed as $\alpha_c = (0, 4, 2)$. Likewise, we specify the other demand vectors. In each case we list the goods demanded in the form

$$\alpha = (\text{crude oil, gasoline, electricity}).$$

The demand vectors are

$$
\begin{aligned}
\text{crude oil industry:} \quad & \alpha_c = (0, 4, 2), \\
\text{refining industry:} \quad & \alpha_r = (8, 0, 6), \\
\text{utility industry:} \quad & \alpha_u = (1, 6, 0), \\
\text{general public:} \quad & \alpha_1 = (1, 9, 5), \\
\text{government:} \quad & \alpha_2 = (8, 8, 8), \\
\text{export firms:} \quad & \alpha_3 = (7, 2, 0).
\end{aligned}
$$

The total demand on all the industries is then

$$\alpha_{\text{total}} = \alpha_c + \alpha_r + \alpha_u + \alpha_1 + \alpha_2 + \alpha_3 = (25, 29, 21).$$

Now suppose that the price of crude oil is \$4 per unit, the price of gasoline \$3 per unit, and the price of electricity \$2 per unit. This can be expressed as a vector $\beta = (4, 3, 2)$. Assuming that the industries produce exactly what is demanded of them, the income of the crude oil industry is 25 units times \$4 which equals \$100. The crude oil industry had to have gasoline and electricity to operate; therefore, its costs were

$$\alpha_c \cdot \beta = (0, 4, 2) \cdot (4, 3, 2) = \$16.$$

Hence, the profit in the crude oil industry is

$$\$100 - \$16 = \$84.$$

Finding the profits (or losses) of the other industries is left as an exercise.

EXERCISES

1. Let $\alpha = (2, -3, 0, 4)$ and let $\beta = (2, -1, 3, x)$.
   (a) Find $\alpha + \beta$.   (b) Find 3α.
   (c) Is $\beta \geq \alpha$? Why?   (d) Find $2\alpha - 3\beta$.
   (e) Find $2\alpha + 3\beta$.   (f) Let $0 = (0, 0, 0, 0)$; find $\alpha + 0$.

2. Represent each of the following vectors on a separate graph, and find the magnitude of each vector.

$$\alpha = (2, -4); \qquad \beta = (2, 4, 9); \qquad \gamma = (4, 0, -1).$$

3. Find the sum and the dot product, if possible, of each of the following pairs of vectors; if that is not possible, state why it is not possible.
   (a) $(1, 2, 3)$, $(-2, 0, 2)$;
   (b) $(6, 0, 4, -1)$, $(3, 1, 0, 2)$;
   (c) $(3, 1, 2)$, $(2, 3)$.

4. In Exercise 3 subtract the second vector from the first vector of each pair if possible. Then find the magnitude of the result in each case, if possible.

5. Find the cosine of the angle between $\alpha = (-2, 2)$ and $\beta = (4, 3)$. What is the significance of the negative answer? Draw a graph.

6. Write an expression for $\alpha^2 = \alpha \cdot \alpha$, where α is an n-dimensional vector. Can α^3 be defined in a similar manner? Why?

7. In two-dimensional space prove the following theorem: If $\alpha \cdot \beta = 0$ and $\alpha \neq 0$ and $\beta \neq 0$, then α is perpendicular to β.

8. A force of 10 pounds pulls due north on an object while a force of $10\sqrt{3}$ pounds pulls due east on the same object. What is the magnitude of the resultant force acting on the object and in what direction does it act?

9. A helicopter pilot wishes to move due north. He has an airspeed of 50 miles per hour. A wind of $25 \sqrt{2}$ miles per hour is blowing from the southeast. In what direction should the pilot point the helicopter, and with what speed will he move?

10. Find the work done by moving an object along a distance given by the vector $(2, 6)$ if the force acting on the particle is $(1, 2)$.

11. Find the work done in moving an object by a force $(2, 2)$ along a distance given by a vector $(4, -1)$.

12. In Example 6 of this section suppose that the same quantity of stock had been purchased a year previously at a market value of $30, $40, $40, $20 per share, respectively. Evaluate the profit (or loss) using vector notation.

13. In Example 8 of this section, find the profits (or losses) of the refining industry and the utility industry.

14. A *unit vector* is a vector with a magnitude equal to one unit. Find a unit vector having the same direction as $(2, 3, 5)$.

NEW VOCABULARY

§1.2	matrix	§1.3	vector
§1.2	zero matrix	§1.3	components
§1.2	null matrix	§1.3	n-dimensional vectors
§1.2	real matrix	§1.3	zero vector
§1.2	address	§1.3	null vector
§1.2	order of a matrix	§1.3	real vector
§1.2	square matrix of order n	§1.3	equal vectors
§1.2	abbreviated notation	§1.3	inequality of real vectors
§1.2	main diagonal	§1.3	addition of vectors
§1.2	submatrix	§1.3	vector multiplied by a scalar
§1.2	partitioned matrix		
§1.2	equal matrices	§1.3	dot product
§1.2	inequality of real ma-trices	§1.3	inner product
		§1.3	scalar product
§1.2	coefficient matrix	§1.3	magnitude
§1.2	augmented matrix	§1.3	unit vector
§1.3	row matrix	§1.3	probability vector
§1.3	column matrix		

Binary Operations on Matrices

2.1 BINARY OPERATIONS

The reader is familiar with the four binary operations of addition, subtraction, multiplication, and division as they are defined and used for real numbers; for example, $5 + 3 = 8$, $5 - 2 = 3$, $4 \cdot 2 = 8$, and $6 \div 2 = 3$. When a new set of elements is introduced, new binary operations may be defined and used; for example, in Section 1.3, we defined the following operations for vectors:

$$(1, 2, 3) + (1, 0, 3) = (2, 2, 6), \qquad (1, 2, 3) \cdot (1, 0, 3) = 10.$$

It should be obvious that there may exist many other sets with associated operations. We are therefore led to a brief discussion of operations in general. We use a small circle "∘" to designate an unspecified operation used to combine two elements.

Definition 2.1. *Let R be a set and let P be a set of ordered pairs of elements (a, b). A **binary operation** "∘" from set P to set R is a correspondence by which each element of P is assigned a unique element of R.*

In this definition if a is an arbitrary element of a set S_1 and b is an arbitrary element of a set S_2, then clearly the operation "∘" is a mapping of the cartesian products $S_1 \times S_2$ into R. If $S_1 = S_2 = S$ and if R is a subset of S so that "∘" is a mapping of $S \times S$ into S, we say that the *set S is* **closed** *under the operation* "∘" or that "∘" is a *binary operation* **over** *S*.

The result of combining a pair of elements a and b by performing the operation "∘" is written $a \circ b$. The reader should be careful to distinguish between the operation and the result of performing it.

Example 1. $\quad 4 + 3 = 7$.

The elements are from the set of integers S, the operation is one known as "addition," and the result of the operation is the sum. Moreover, the set of integers is closed under this operation. We say that addition is a binary operation over S.

Example 2. $(3, 2, 1) \cdot (5, -4, 0) = 7.$

The elements of the set are real 3-dimensional vectors, the operation is a type of vector multiplication, and the result is called the scalar product. The resulting element is a scalar rather than a vector; thus, the set of vectors is not closed under this type of vector multiplication. We have an illustration of a binary operation from the set of ordered pairs of real 3-dimensional vectors to the set of real numbers.

It is the purpose of this chapter to discuss certain binary operations associated with sets of matrices, to define these operations, and then to use the results of performing them.

EXERCISES

1. Let S be the set of integers.
   (a) Is subtraction a binary operation over S?
   (b) Is division a binary operation over S?

2. Is the set of rational numbers closed under the binary operation of subtraction? What number must be excluded from the set in order to have a subset that is closed under division? What is the result of performing the operation of division called?

3. Suppose we define a binary operation (∗) over the set of positive integers as $a * b = a + 2b$. (For example, $4 * 3 = 4 + 2(3) = 10$). Does $x * y = y * x$?

2.2 MATRIX ADDITION

Two matrices of the same order can be added according to the following definition.

Definition 2.2.   *Given two matrices $A = [a_{ij}]_{(m,n)}$ and $B = [b_{ij}]_{(m,n)}$, their sum is defined to be $A + B = [a_{ij} + b_{ij}]_{(m,n)}$.*

When two matrices are of the same order, they are said to be *conformable for addition*.

Example 1.

$$(a)\begin{bmatrix} 2 & 2 \\ 6 & 9 \end{bmatrix} + \begin{bmatrix} 3 & 1 \\ 1 & 0 \end{bmatrix} = \begin{bmatrix} (2+3) & (2+1) \\ (6+1) & (9+0) \end{bmatrix} = \begin{bmatrix} 5 & 3 \\ 7 & 9 \end{bmatrix}.$$

(b) $\begin{bmatrix} 2 & 1 \\ 4 & 0 \end{bmatrix} + \begin{bmatrix} 3 & 1 & 2 \\ 1 & 0 & 6 \end{bmatrix}$ is undefined because the two matrices are not conformable for addition.

Theorem 2.1. *Commutative Property.* *If two matrices A and B with scalar entries are conformable for addition then*

$$A + B = B + A.$$

Proof. The proof of this theorem is left as an exercise.

Theorem 2.2. *Associative Property.* *If three matrices, A, B, and C with scalar entries are of the same order then*

$$A + (B + C) = (A + B) + C.$$

Proof.

STATEMENT	REASON
(1) $[a_{ij}]_{(m,n)} + ([b_{ij}]_{(m,n)} + [c_{ij}]_{(m,n)})$ $= [a_{ij}]_{(m,n)} + [b_{ij} + c_{ij}]_{(m,n)}$	(1) By definition of matrix addition.
(2) $= [a_{ij} + (b_{ij} + c_{ij})]_{(m,n)}$	(2) By matrix addition.
(3) $= [(a_{ij} + b_{ij}) + c_{ij}]_{(m,n)}$	(3) Addition of scalars is associative.
(4) $= [a_{ij} + b_{ij}]_{(m,n)} + [c_{ij}]_{(m,n)}$	(4) By matrix addition.
(5) $= ([a_{ij}]_{(m,n)} + [b_{ij}]_{(m,n)}) + [c_{ij}]_{(m,n)}.$	(5) By matrix addition.

Theorem 2.3. *Cancellation Property.* *If three matrices A, B, and C with scalar entries are of the same order then*

$$A + B = A + C \;\Rightarrow\; B = C,$$

and

$$A + B = C + B \;\Rightarrow\; A = C,$$

where the symbol "⇒" means "implies that."

Proof. The proof is left as an exercise.

APPLICATIONS

Example 2. Assume that a manufacturer produces a certain alloy. The costs of purchasing and transporting specific amounts of three necessary raw materials from two different locations are given, respectively, by the following matrices:

$$A = \begin{bmatrix} 16 & 20 \\ 10 & 16 \\ 9 & 4 \end{bmatrix} \begin{matrix} \text{ORE } R \\ \text{ORE } \bar{S} \\ \text{ORE } T \end{matrix},$$

with columns PURCHASE COST and TRANSPORTATION COST,

$$B = \begin{bmatrix} 12 & 10 \\ 14 & 14 \\ 12 & 10 \end{bmatrix} \begin{matrix} \text{ORE } R \\ \text{ORE } S \\ \text{ORE } T \end{matrix}.$$

The matrix representing the total purchase and transportation costs of each type of ore from both locations is

$$A + B = \begin{bmatrix} 28 & 30 \\ 24 & 30 \\ 21 & 14 \end{bmatrix}.$$

Example 3. An up-to-date inventory is a vital necessity in many businesses; if the process can be quantified, computer help becomes available for keeping an up-to-date record. Suppose that a district sales manager for a certain make of automobile wishes to keep a day-by-day inventory of the colors and models he has on hand. He requires that each retailer keep and report an inventory matrix of the following form:

	TAN	RED	WHITE	GREEN	BLUE	
$A_i =$	a_{11}	a_{12}	a_{13}	a_{14}	a_{15}	CONVERTIBLE
	a_{21}	a_{22}	a_{23}	a_{24}	a_{25}	2-DOOR SEDAN .
	a_{31}	a_{32}	a_{33}	a_{34}	a_{35}	4-DOOR SEDAN

If this sales manager has 99 retailers in his district, then the total inventory matrix T for this district is

$$A_1 + A_2 + \cdots + A_{99} = T.$$

This can be found very simply by programming a computer to do the necessary arithmetic calculations. Those items which are in total short supply can then be ordered.

EXERCISES

1. Calculate, if possible, each of the following:

(a) $\begin{bmatrix} 1 & 2 \\ 3 & 4 \end{bmatrix} + \begin{bmatrix} 3 & 2 \\ 0 & 1 \end{bmatrix}$; (b) $\begin{bmatrix} 1 & 2 \\ 3 & 4 \end{bmatrix} + \begin{bmatrix} 6 \\ 4 \end{bmatrix}$;

(c) $\begin{bmatrix} 3 & 4 \\ \sqrt{2} & 1 \\ 4 & 0 \end{bmatrix} + \begin{bmatrix} 0 & 0 \\ 1 & -1 \\ 2 & i \end{bmatrix}$; (d) $\begin{bmatrix} 3 \\ 4 \\ 6 \end{bmatrix} + \begin{bmatrix} 2 \\ 1 \\ 4 \end{bmatrix} + \begin{bmatrix} 0 \\ 0 \\ 0 \end{bmatrix}$.

2. If A is a 2 by 3 matrix and B is a 3 by 2 matrix, are A and B conformable for addition?

3. Prove Theorem 2.1.

4. Prove Theorem 2.3.

2.3 MULTIPLICATION BY A SCALAR

Consider the sum $A + A + A$. Can we say that this sum is equal to $3A$? An affirmative answer is justified if a matrix multiplied by a scalar is defined in the following manner:

Definition 2.3. *Given a matrix* $A = [a_{ij}]_{(m,n)}$ *and a scalar* c, *then* $cA = [ca_{ij}]_{(m,n)}$ *and* $Ac = [a_{ij}c]_{(m,n)}$.

Thus, a matrix is multiplied by a scalar by multiplying every entry of the matrix by that scalar. If c belongs to the set of all scalars S and if A belongs to the set of all matrices M, then this kind of multiplication is a binary operation from the set of all ordered pairs (c, A) or (A, c) to M. Making use of this concept, *subtraction* of matrices can be defined as

$$A - B = A + (-1)B.$$

Example 1.

$$2\begin{bmatrix} 3 & 1 \\ 4 & 3 \end{bmatrix} - \begin{bmatrix} 1 & 6 \\ 0 & 4 \end{bmatrix} = 2\begin{bmatrix} 3 & 1 \\ 4 & 3 \end{bmatrix} + (-1)\begin{bmatrix} 1 & 6 \\ 0 & 4 \end{bmatrix}$$

$$= \begin{bmatrix} 6 & 2 \\ 8 & 6 \end{bmatrix} + \begin{bmatrix} -1 & -6 \\ 0 & -4 \end{bmatrix} = \begin{bmatrix} 5 & -4 \\ 8 & 2 \end{bmatrix}.$$

APPLICATIONS

Example 2.   In Example 3 of Section 2.2 the sales manager could calculate his inventory, T_2, after any given week by subtracting what was sold, S_2, that week from the total inventory matrix after the previous week, T_1; that is,

$$T_2 = T_1 - S_2 = T_1 + (-1)S_2.$$

However, in order to stay well ahead of the demand, suppose he orders and receives a shipment of new cars which is exactly double what he sold, S_1, the first week.   Here we have

$$T_2 = T_1 - S_2 + 2S_1.$$

★ **Example 3.**   In Chapter 8, we discuss certain important engineering problems which require consideration of the matrix $\lambda I - A$, where A is a square matrix, λ is a scalar, and I is a square matrix in which every entry of the main diagonal is 1 and every other entry is 0.   The matrix I is the very important *identity matrix*. Its name is derived from the fact that it is the *identity element* for multiplication for the set of n by n matrices; that is, $AI = IA = A$. A matrix λI is known as a *scalar matrix*.

Thus, if

$$A = \begin{bmatrix} 6 & 3 & 4 \\ 3 & 2 & 0 \\ 4 & 0 & 1 \end{bmatrix},$$

then

$$\lambda I - A = \lambda \begin{bmatrix} 1 & 0 & 0 \\ 0 & 1 & 0 \\ 0 & 0 & 1 \end{bmatrix} + (-1) \begin{bmatrix} 6 & 3 & 4 \\ 3 & 2 & 0 \\ 4 & 0 & 1 \end{bmatrix}$$

$$= \begin{bmatrix} \lambda & 0 & 0 \\ 0 & \lambda & 0 \\ 0 & 0 & \lambda \end{bmatrix} + \begin{bmatrix} -6 & -3 & -4 \\ -3 & -2 & 0 \\ -4 & 0 & -1 \end{bmatrix}$$

$$= \begin{bmatrix} \lambda-6 & -3 & -4 \\ -3 & \lambda-2 & 0 \\ -4 & 0 & \lambda-1 \end{bmatrix}.$$

Example 4.   Consider the set of all 2 by 2 matrices of the form

$$aI + bJ,$$

where a and b are real numbers and

$$J = \begin{bmatrix} 0 & 1 \\ -1 & 0 \end{bmatrix}.$$

Consider the sum of any two matrices of this set, say

$$(aI + bJ) + (cI + dJ) = (a + c)I + (b + d)J.$$

Notice the correspondence between this sum and the sum of complex numbers $(a + bi) + (c + di) = (a + c) + (b + d)i$. It is shown after Section 2.4 that this correspondence is also preserved under multiplication.

EXERCISES

1. Given $A = \begin{bmatrix} 2 & -1 \\ -3 & -4 \end{bmatrix}$ and $B = \begin{bmatrix} -2 & 0 \\ -1 & 3 \end{bmatrix}$, calculate:

    (a) $3A$;   (b) $-2B$;   (c) $-A$;   (d) $A + 3B$;   (e) $\frac{1}{2}B - 2A$.
    (f) Find C if $B + C = A$.   (g) Find D if $A - 2D = 2B$.

2. Let $A = \begin{bmatrix} 5 & 10 & 20 \\ -65 & 15 & -10 \end{bmatrix}$. Find a matrix B which is a scalar multi-

ple of A and which has 2 as its entry in the first row and second column.

3. Two matrices can be added only when they are of the same order. Is there any restriction on the multiplication of a matrix by a scalar?

4. Find $\lambda I - A$ if $A = \begin{bmatrix} 3 & 2 \\ 1 & 6 \end{bmatrix}$. (See Example 3 of this section.)

5. What is a scalar matrix? Illustrate with a 4 by 4 matrix.

6. Write the matrices which correspond to the complex numbers $4 + 2i$ and $2 - i$ as shown in Example 4 of this section. Demonstrate that this correspondence is preserved under addition.

7. Prove that $Ac = cA$, where A is a matrix and c is a scalar.

2.4 MATRIX MULTIPLICATION

Now that we have learned to add and subtract two matrices, it is logical to inquire about another binary operation, — multiplication. Matrix multiplication is now defined in such a way that the product of a row matrix of n coefficients by a column matrix of n variables is a 1 by 1 matrix whose single entry is a linear function of those n variables.

$$[a_{11} \quad a_{12} \quad \cdots \quad a_{1n}] \begin{bmatrix} x_1 \\ x_2 \\ \cdot \\ \cdot \\ \cdot \\ x_n \end{bmatrix} = [a_{11}x_1 + a_{12}x_2 + \cdots + a_{1n}x_n].$$

Notice that the single entry of the resulting matrix is similar to the dot product of two vectors. Also, we define the product of an m by n matrix of coefficients by an n-rowed column matrix of variables to be an m by 1 column matrix, each of whose entries is a linear function of the n variables.

$$\begin{bmatrix} a_{11} & a_{12} & \cdots & a_{1n} \\ a_{21} & a_{22} & \cdots & a_{2n} \\ \cdot & \cdot & & \cdot \\ \cdot & \cdot & & \cdot \\ \cdot & \cdot & & \cdot \\ a_{m1} & a_{m2} & \cdots & a_{mn} \end{bmatrix} \begin{bmatrix} x_1 \\ x_2 \\ \cdot \\ \cdot \\ \cdot \\ x_n \end{bmatrix} = \begin{bmatrix} (a_{11}x_1 + a_{12}x_2 + \cdots + a_{1n}x_n) \\ (a_{21}x_1 + a_{22}x_2 + \cdots + a_{2n}x_n) \\ \cdot \\ \cdot \\ \cdot \\ (a_{m1}x_1 + a_{m2}x_2 + \cdots + a_{mn}x_n) \end{bmatrix}.$$

Note that this result is a column matrix whose entries are dot products. These ideas can be generalized and are stated formally in the following definition.

Definition 2.4. *Let A be an m by p matrix and B be a p by n matrix. The* **product** *$C = AB$ is an m by n matrix, where each entry c_{ij} of C is obtained by multiplying corresponding entries of the ith row of A by those of the jth column of B and then adding the results.*

In Definition 2.4, if A and B are real matrices, then each c_{ij} is simply the dot product of the ith row of A and the jth column of B. The following diagrams illustrate Definition 2.4.

$$\begin{bmatrix} a_{11} & a_{12} & \cdots & a_{1p} \\ a_{21} & a_{22} & \cdots & a_{2p} \\ \cdot & \cdot & \cdot & \cdot \\ \cdot & \cdot & \cdot & \cdot \\ \cdot & \cdot & \cdot & \cdot \\ a_{m1} & a_{m2} & \cdots & a_{mp} \end{bmatrix} \begin{bmatrix} b_{11} & b_{12} & \cdots & b_{1n} \\ b_{21} & b_{22} & \cdots & b_{2n} \\ \cdot & \cdot & \cdot & \cdot \\ \cdot & \cdot & \cdot & \cdot \\ \cdot & \cdot & \cdot & \cdot \\ b_{p1} & b_{p2} & \cdots & b_{pn} \end{bmatrix} = \begin{bmatrix} c_{11} & c_{12} & \cdots & c_{1n} \\ c_{21} & c_{22} & \cdots & c_{2n} \\ \cdot & \cdot & \cdot & \cdot \\ \cdot & \cdot & \cdot & \cdot \\ \cdot & \cdot & \cdot & \cdot \\ c_{m1} & c_{m2} & \cdots & c_{mn} \end{bmatrix},$$

where

$$c_{11} = a_{11}b_{11} + a_{12}b_{21} + \cdots + a_{1p}b_{p1}.$$

$$\begin{bmatrix} a_{11} & \cdots & a_{1p} \\ \cdot & & \cdot \\ \cdot & & \cdot \\ \cdot & & \cdot \\ \boxed{a_{i1}} & \cdots & \boxed{a_{ip}} \\ \cdot & & \cdot \\ \cdot & & \cdot \\ \cdot & & \cdot \\ a_{m1} & \cdots & a_{mp} \end{bmatrix} \begin{bmatrix} b_{11} & \cdots & \boxed{b_{1j}} & \cdots & b_{1n} \\ \cdot & & \cdot & & \\ \cdot & & \cdot & & \\ \cdot & & \cdot & & \\ \cdot & & \cdot & & \cdot \\ \cdot & & \cdot & & \\ \cdot & & \cdot & & \\ \cdot & & \cdot & & \\ b_{p1} & \cdots & \boxed{b_{pj}} & \cdots & b_{pn} \end{bmatrix} = \begin{bmatrix} c_{11} & \cdots & c_{1n} \\ \cdot & & \cdot \\ \cdot & & \cdot \\ \cdot & \boxed{c_{ij}} & \cdot \\ \cdot & & \cdot \\ \cdot & & \cdot \\ c_{m1} & \cdots & c_{mn} \end{bmatrix},$$

where

$$c_{ij} = a_{i1}b_{1j} + a_{i2}b_{2j} + \cdots + a_{ip}b_{pj}.$$

Example 1. Let

$$A = \begin{bmatrix} 1 & 2 \\ 1 & 0 \end{bmatrix}, \quad B = \begin{bmatrix} 3 & 0 & 1 \\ 0 & 1 & 1 \end{bmatrix},$$

then

$$AB = \begin{bmatrix} 1 & 2 \\ 1 & 0 \end{bmatrix}\begin{bmatrix} 3 & 0 & 1 \\ 0 & 1 & 1 \end{bmatrix}.$$

Using the terminology of the definition we note that here $m = 2$, $p = 2$, and $n = 3$.

$$AB = \begin{bmatrix} (1\cdot3 + 2\cdot0) & (1\cdot0 + 2\cdot1) & (1\cdot1 + 2\cdot1) \\ (1\cdot3 + 0\cdot0) & (1\cdot0 + 0\cdot1) & (1\cdot1 + 0\cdot1) \end{bmatrix}$$

$$= \begin{bmatrix} 3 & 2 & 3 \\ 3 & 0 & 1 \end{bmatrix}.$$

In the preceding example it should be noted that the product BA does not exist. By the definition, it is apparent that matrices can be multiplied only when the number of columns of the left matrix is equal to the number of rows of the right matrix; when this is the case we say that the left matrix is *conformable for multiplication* to the right matrix.

Example 2.

$$AB = \begin{bmatrix} 3 & 2 \\ 1 & 4 \end{bmatrix}\begin{bmatrix} 1 \\ 2 \end{bmatrix} = \begin{bmatrix} (3\cdot1 + 2\cdot2) \\ (1\cdot1 + 4\cdot2) \end{bmatrix} = \begin{bmatrix} 7 \\ 9 \end{bmatrix},$$

but B is not conformable to A for multiplication; hence BA does not exist. Consider the products of two other matrices,

$$CD = \begin{bmatrix} 1 & 2 & 3 \end{bmatrix} \begin{bmatrix} 0 \\ 5 \\ 4 \end{bmatrix} = [1 \cdot 0 + 2 \cdot 5 + 3 \cdot 4] = [22];$$

$$DC = \begin{bmatrix} 0 \\ 5 \\ 4 \end{bmatrix} \begin{bmatrix} 1 & 2 & 3 \end{bmatrix} = \begin{bmatrix} 0 \cdot 1 & 0 \cdot 2 & 0 \cdot 3 \\ 5 \cdot 1 & 5 \cdot 2 & 5 \cdot 3 \\ 4 \cdot 1 & 4 \cdot 2 & 4 \cdot 3 \end{bmatrix} = \begin{bmatrix} 0 & 0 & 0 \\ 5 & 10 & 15 \\ 4 & 8 & 12 \end{bmatrix}.$$

The previous two examples make it quite obvious that in general $AB \neq BA$; that is, the commutative property for matrix multiplication is *not* valid. Also, the cancellation property that *was* valid for matrix addition is *not* valid for matrix multiplication; that is, $AB = AC$ does *not* imply that $B = C$. The following two properties, however, are valid; their proofs are simplified if sigma notation is used. A discussion of the sigma notation is given in Sections A.1 and A.2.

Theorem 2.4. Associative Property. *Given that the three matrices with scalar entries A, B, and C are conformable for multiplication, then*

$$A(BC) = (AB)C.$$

Proof. See Theorem A.2, page 154.

Theorem 2.5. Distributive Property. *Assuming conformability, and that A, B, and C are matrices with scalar entries, then*

$$A(B + C) = AB + AC,$$

and

$$(A + B)C = AC + BC.$$

Proof. See Theorem A.3, page 155.

Example 3. Given

$$A = \begin{bmatrix} 3 & 2 \\ 4 & 0 \end{bmatrix}, \quad B = \begin{bmatrix} 1 & 2 \\ 0 & 1 \end{bmatrix}, \quad C = \begin{bmatrix} 0 & 2 \\ 1 & 3 \end{bmatrix}.$$

$$A(BC) = \begin{bmatrix} 3 & 2 \\ 4 & 0 \end{bmatrix} \begin{bmatrix} 2 & 8 \\ 1 & 3 \end{bmatrix} = \begin{bmatrix} 8 & 30 \\ 8 & 32 \end{bmatrix}.$$

$$(AB)C = \begin{bmatrix} 3 & 8 \\ 4 & 8 \end{bmatrix} \begin{bmatrix} 0 & 2 \\ 1 & 3 \end{bmatrix} = \begin{bmatrix} 8 & 30 \\ 8 & 32 \end{bmatrix}.$$

$$A(B + C) = \begin{bmatrix} 3 & 2 \\ 4 & 0 \end{bmatrix} \left(\begin{bmatrix} 1 & 2 \\ 0 & 1 \end{bmatrix} + \begin{bmatrix} 0 & 2 \\ 1 & 3 \end{bmatrix} \right) = \begin{bmatrix} 5 & 20 \\ 4 & 16 \end{bmatrix}.$$

$$AB + AC = \begin{bmatrix} 3 & 2 \\ 4 & 0 \end{bmatrix} \begin{bmatrix} 1 & 2 \\ 0 & 1 \end{bmatrix} + \begin{bmatrix} 3 & 2 \\ 4 & 0 \end{bmatrix} \begin{bmatrix} 0 & 2 \\ 1 & 3 \end{bmatrix} = \begin{bmatrix} 5 & 20 \\ 4 & 16 \end{bmatrix}.$$

For reference purposes a list of additional properties is given. The proofs are left as exercises for the reader.

Theorem 2.6. *Assuming conformability, and that A and B are matrices with scalar entries and c and d are scalars, the following identities hold:*

$$cA = Ac,$$
$$c(dA) = (cd)A,$$
$$c(AB) = (cA)B,$$
$$A(cB) = (Ac)B,$$
$$(c + d)A = cA + dA,$$
$$(A + B)c = Ac + Bc.$$

Positive integral powers of square matrices are defined as they are for scalars

$$A^2 = AA, \qquad A^3 = AAA, \qquad A^4 = AAAA, \qquad \text{etc.}$$

APPLICATIONS

★ **Example 4.** One of the most useful applications of matrices is the matrix representation of a system of linear equations. The linear system

$$\begin{cases} a_{11}x_1 + a_{12}x_2 + \cdots + a_{1n}x_n = b_1, \\ a_{21}x_1 + a_{22}x_2 + \cdots + a_{2n}x_n = b_2, \\ \phantom{a_{21}x_1} \vdots \\ a_{m1}x_1 + a_{m2}x_2 + \cdots + a_{mn}x_n = b_m, \end{cases}$$

can be expressed as

$$AX = B,$$

where

$$A = \begin{bmatrix} a_{11} & a_{12} & \cdots & a_{1n} \\ a_{21} & a_{22} & \cdots & a_{2n} \\ \cdot & \cdot & \cdot & \cdot \\ \cdot & \cdot & \cdot & \cdot \\ \cdot & \cdot & \cdot & \cdot \\ a_{m1} & a_{m2} & \cdots & a_{mn} \end{bmatrix}, \quad X = \begin{bmatrix} x_1 \\ x_2 \\ \cdot \\ \cdot \\ \cdot \\ x_n \end{bmatrix}, \quad B = \begin{bmatrix} b_1 \\ b_2 \\ \cdot \\ \cdot \\ \cdot \\ b_m \end{bmatrix}.$$

Specifically, the matrix equation

$$\begin{bmatrix} 2 & 4 \\ 1 & 3 \end{bmatrix}\begin{bmatrix} x_1 \\ x_2 \end{bmatrix} = \begin{bmatrix} 3 \\ 1 \end{bmatrix},$$

or by matrix multiplication

$$\begin{bmatrix} (2x_1 + 4x_2) \\ (x_1 + 3x_2) \end{bmatrix} = \begin{bmatrix} 3 \\ 1 \end{bmatrix},$$

represents

$$\begin{cases} 2x_1 + 4x_2 = 3, \\ x_1 + 3x_2 = 1, \end{cases}$$

because corresponding entries of equal matrices are equal.

Example 5. Suppose that a certain fruit packer in Florida has a boxcar loaded with fruit ready to be shipped north. The load consists of 900 boxes of oranges, 700 boxes of grapefruit, and 400 boxes of tangerines. The market prices, per box, of the different types of fruit in the various cities are given by the following chart.

	oranges	grapefruit	tangerines
New York	$4	$2	$3
Cleveland	$5	$1	$2
St. Louis	$4	$3	$2
Oklahoma City	$3	$2	$5

To which city should the carload of fruit be sent for the packer to get maximum gross receipts for his fruit?

Solution: Consider the chart above as the "price matrix," and form the

"quantity matrix" $\begin{bmatrix} 900 \\ 700 \\ 400 \end{bmatrix}$ BOXES BOXES BOXES. The product of these matrices, as shown below, yields an "income matrix," each entry of which represents the total income from the sale of all of the fruit in each city.

$$\begin{bmatrix} 4 & 2 & 3 \\ 5 & 1 & 2 \\ 4 & 3 & 2 \\ 3 & 2 & 5 \end{bmatrix} \begin{bmatrix} 900 \\ 700 \\ 400 \end{bmatrix} = \begin{bmatrix} 3600 + 1400 + 1200 \\ 4500 + 700 + 800 \\ 3600 + 2100 + 800 \\ 2700 + 1400 + 2000 \end{bmatrix} = \begin{bmatrix} 6200 \\ 6000 \\ 6500 \\ 6100 \end{bmatrix}.$$

The largest entry in the income matrix is 6500; therefore, the greatest income comes from St. Louis.

Example 6. Consider a communications system in which some stations may *not* speak to other stations (these communication centers could represent women or nations as well as electronic apparatus). If the ith station can speak to the jth station then define $a_{ij} = 1$; otherwise, $a_{ij} = 0$. (This is similar to Example 5 and Exercise 9 of Section 1.2.) Suppose that we have four stations which communicate according to the following matrix. We assume that a station does not speak to itself directly, hence the entries on the main diagonal are 0.

$$\begin{array}{c} \text{RECEIVERS} \\ \begin{array}{cccc} \#1 & \#2 & \#3 & \#4 \end{array} \\ A = \begin{bmatrix} 0 & 0 & 0 & 1 \\ 1 & 0 & 0 & 1 \\ 1 & 0 & 0 & 0 \\ 0 & 1 & 1 & 0 \end{bmatrix} \begin{array}{c} \#1 \\ \#2 \\ \#3 \\ \#4 \end{array} \end{array} \quad \text{SPEAKERS}$$

This matrix indicates that stations #2 and #4 are the only stations that directly reciprocate communication, whereas stations #2 and #3 do not directly communicate with each other at all.

This question can be raised: Which stations can communicate indirectly with the help of intermediaries? Let us look at the entry in the fourth row and first column of

$$\begin{array}{c} \begin{array}{cccc} \#1 & \#2 & \#3 & \#4 \end{array} \\ A^2 = \begin{bmatrix} 0 & 1 & 1 & 0 \\ 0 & 1 & 1 & 1 \\ 0 & 0 & 0 & 1 \\ 2 & 0 & 0 & 1 \end{bmatrix} \begin{array}{c} \#1 \\ \#2 \\ \#3 \\ \#4 \end{array} ;$$

This entry b_{41} of the matrix $B = A^2$ equals

$$a_{41}a_{11} + a_{42}a_{21} + a_{43}a_{31} + a_{44}a_{41}$$

(these a_{ij} belong to the matrix A). Any one of these four products $a_{4k}a_{k1}$ (where $k = 1, 2, 3,$ or 4) is equal to 1 only if both a_{4k} and a_{k1} are 1, that is, when station #4 can speak to station #k (where k = 1, 2, 3, or 4) and station #k can in turn speak to station #1. Here, we find b_{41} = 2, that is, there are two possible channels of communication by which station #4 can speak to station #1 through an intermediary (or by one relay).

The c_{ij} entry of $C = A^3$ tells how many channels of communication are open between station #i and station #j using two relays, and so forth for A^{n+1}. Furthermore, $A + A^2$ gives the total number of channels of communication that are open between various stations for either zero *or* one relay; generalizing, $A + A^2 + A^3 + \cdots + A^{n+1}$ gives the total number of channels of communication that are open between the various stations, with no more than n relays.

Example 7.   The following equations, involving products of matrices,

$$\sigma_x \sigma_y = -\sigma_y \sigma_x = i\sigma_z,$$
$$\sigma_y \sigma_z = -\sigma_z \sigma_y = i\sigma_x,$$
$$\sigma_z \sigma_x = -\sigma_x \sigma_z = i\sigma_y,$$

where

$$\sigma_x = \begin{bmatrix} 0 & 1 \\ 1 & 0 \end{bmatrix}, \qquad \sigma_y = \begin{bmatrix} 0 & -i \\ i & 0 \end{bmatrix}, \qquad \text{and} \qquad \sigma_z = \begin{bmatrix} 1 & 0 \\ 0 & -1 \end{bmatrix}$$

are very important in atomic physics. The matrices σ_x, σ_y, and σ_z are known as the Pauli spin matrices (introduced by Wolfgang Pauli, Jr., Austrian and American, 1900–) and are used to describe the intrinsic angular momentum of an electron. Furthermore the reader can prove that $\sigma_x^2 = \sigma_y^2 = \sigma_z^2 = I$.

Example 8.   In Example 7 of Section 1.3 a probability vector was defined as a vector with nonnegative components whose sum is 1. A matrix that consists of rows of probability vectors is called a *stochastic matrix;* such matrices are useful in connection with Markov chains. Consider a voting precinct in which the predicted voting probabilities of the people are shown in Figure 2.1. For example, of the people who voted for the Republicans in the previous election, it is predicted that 60% will again vote Republican, while 30% will vote for the Democrats, and 10% will vote for a third party. This information can be stated as a stochastic matrix

$$P = \begin{array}{c} \\ \\ \end{array} \begin{array}{ccc} \text{TO } D & \text{TO } R & \text{TO } T \\ \begin{bmatrix} 0.60 & 0.20 & 0.20 \\ 0.30 & 0.60 & 0.10 \\ 0.30 & 0.20 & 0.50 \end{bmatrix} & \begin{array}{l} \text{FROM } D \\ \text{FROM } R \\ \text{FROM } T \end{array} \end{array}.$$

The matrix P represents a one-stage transition.

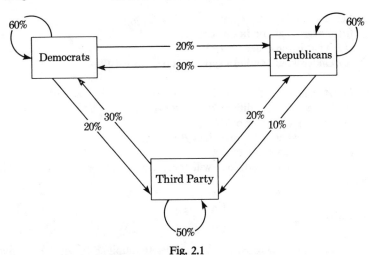

Fig. 2.1

It can be shown that the matrix

$$P^2 = \begin{bmatrix} 0.60 & 0.20 & 0.20 \\ 0.30 & 0.60 & 0.10 \\ 0.30 & 0.20 & 0.50 \end{bmatrix}^2 = \begin{bmatrix} 0.48 & 0.28 & 0.24 \\ 0.39 & 0.44 & 0.17 \\ 0.39 & 0.28 & 0.33 \end{bmatrix}$$

represents a two-stage transition; that is, 48% of the people who voted Democratic in the first election will be expected to vote Democratic in the third election. Note that P^2 is also a stochastic matrix. Moreover, this method of stating predictions can be generalized to P^n, a stochastic matrix which represents the probabilities of an n-stage transition.

EXERCISES

1. Multiply the following, if possible:

(a) $\begin{bmatrix} 2 & 1 \\ 3 & 4 \end{bmatrix} \begin{bmatrix} 0 & 1 \\ 2 & -1 \end{bmatrix}$;

(b) $\begin{bmatrix} 2 & 1 \\ 6 & 0 \end{bmatrix} \begin{bmatrix} -1 \\ 4 \end{bmatrix}$;

(c) $[2 \quad 1 \quad 0] \begin{bmatrix} 4 & 0 \\ 0 & 2 \\ -1 & 1 \end{bmatrix}$;

(d) $\begin{bmatrix} 4 & 2 \\ 3 & 1 \end{bmatrix} \begin{bmatrix} 1 & 0 \\ 0 & 1 \end{bmatrix}$;

(e) $\begin{bmatrix} 9 & 6 & 2 \\ 4 & 3 & 1 \end{bmatrix} \begin{bmatrix} 2 & 4 \\ 0 & 2 \end{bmatrix}$;

(f) $\begin{bmatrix} 2 \\ 3 \end{bmatrix} [3 \quad -1]$.

2. Let $A = [a_{ij}]_{(3, t)}$ and $B = [b_{ij}]_{(4, 5)}$.
   (a) Under what conditions does AB exist?
   (b) What is the order of AB?
   (c) Under what conditions, if any, does BA exist?

3. Let $A = [a_{ij}]_{(m, n)}$ and $B = [b_{ij}]_{(r, t)}$.
   (a) Under what conditions does AB exist?
   (b) What is the order of AB?
   (c) Under what conditions does BA exist?
   (d) What is the order of BA?
   (e) Under what conditions will the order of AB be the same as that
of BA?

4. Let $A = \begin{bmatrix} 2 & 0 \\ 3 & 1 \end{bmatrix}$, $B = \begin{bmatrix} 4 & -1 \\ 0 & 2 \end{bmatrix}$, $I = \begin{bmatrix} 1 & 0 \\ 0 & 1 \end{bmatrix}$, $0 = \begin{bmatrix} 0 & 0 \\ 0 & 0 \end{bmatrix}$.

   (a) Premultiply B by A, that is,   (b) Postmultiply B by A, that is, find
        find AB. BA.
   (c) Find B^2.   (d) Find B^3.
   (e) Find IB.   (f) Find $0B$.
   (g) Find I^3.

5. Let $A = [a_{ij}]_{(m, p)}$. Under what conditions does A^n exist?

6. Prove the six identities of Theorem 2.6:
   (a) $cA = Ac$;   (b) $c(dA) = (cd)A$;
   (c) $c(AB) = (cA)B$;   (d) $A(cB) = (Ac)B$;
   (e) $(c + d)A = cA + dA$;   (f) $(A + B)c = Ac + Bc$.

7. Express the linear system

$$\begin{cases} x_1 + x_2 + x_3 = 4, \\ x_1 - x_2 + 2x_3 = 9, \\ 2x_1 \quad\quad + x_3 = 6, \end{cases}$$

in matrix form and state what each matrix corresponds to.

8. Write out the system represented by

$$\begin{bmatrix} 2 & 0 \\ 1 & 3 \\ 4 & 2 \end{bmatrix} \begin{bmatrix} x_1 \\ x_2 \end{bmatrix} = \begin{bmatrix} 2 \\ 1 \\ 3 \end{bmatrix}.$$

√ **9.** Let the matrix $A = \begin{bmatrix} 2 & 1 \\ 4 & 3 \end{bmatrix}$ represent the number of gadgets R and S
that factories P and Q can produce in a day, according to the table below.

	Factory P	Factory Q
Gadget R	2 per day	1 per day
Gadget S	4 per day	3 per day

Let $N = \begin{bmatrix} 5 \\ 6 \end{bmatrix}$ represent the number of days the two factories operate, that is,

P operates 5 days per week and Q operates 6 days per week. Find AN and
state what it represents.

10. In Exercise 9 suppose $N = \begin{bmatrix} x_1 \\ x_2 \end{bmatrix}$. What is the interpretation of

$$A \begin{bmatrix} x_1 \\ x_2 \end{bmatrix} \geq \begin{bmatrix} 9 \\ 8 \end{bmatrix} ?$$

√ **11.** (a) Suppose that four legislators influence each other according to Figure 2.2. Write the matrix that shows the number of ways in which any one legislator can influence another using at most one relay.

(b) Rank the legislators according to the total number of influence channels that each can exert using at most one relay.

12. In Example 7 of this section, justify each of the stated equations, including

$$\sigma_x^2 = \sigma_y^2 = \sigma_z^2 = I.$$

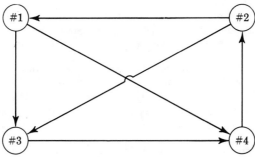

Fig. 2.2

13. Draw a diagram illustrating the communication channels represented by matrix A in Example 6 of this section, and verify visually the conclusions made concerning $A + A^2$.

2.5 OTHER BINARY OPERATIONS (*OPTIONAL*)

Two binary operations that have proved useful in connection with switching functions in electrical engineering are discussed next. First, however, we must define what is meant by a Boolean algebra (named for George Boole, English, 1815–1864) and a Boolean matrix.

Definition 2.5. *A **Boolean algebra** consists of a set of elements S, two operations $\cup$ and $\cap$, and the following postulates;*
(1) *$\cup$ and $\cap$ are commutative on S.*
(2) *$\cup$ is distributive with respect to $\cap$ on S, and $\cap$ is distributive with respect to $\cup$ on S.*
(3) *S contains identity elements 0 for $\cup$, and 1 for $\cap$.*
(4) *For every element x in S there is an element designated x', also in S, such that*

$$x \cup x' = 1 \qquad and \qquad x \cap x' = 0.$$

Thus, if x, y, and z are elements of S:

(1) $x \cup y = y \cup x$; $x \cap y = y \cap x$.
(2) $x \cup (y \cap z)$ $x \cap (y \cup z)$
   $\quad = (x \cup y) \cap (x \cup z)$; $\quad = (x \cap y) \cup (x \cap z)$.
(3) $x \cup 0 = x$; $x \cap 1 = x$.

Example 1. The set of all subsets of a given set U, with $\cup$ representing the union of two sets and $\cap$ representing the intersection of two sets, forms a Boolean algebra. The null set ϕ is the identity element for $\cup$, and the set U is the identity element for $\cap$.

Example 2. If we let the set S consist of only two elements 0 and 1 and define $\cup$ and $\cap$ as we did in Example 1 of Section 1.1 we find another example of a Boolean algebra (the reader should review that example). Because there are only two elements, this structure is known as a *binary Boolean algebra*.

Definition 2.6. *Matrices whose entries are elements of a Boolean algebra are called **Boolean matrices**.*

Definition 2.7. *For two m by n Boolean matrices A and B,*

$$A \cup B = [(a_{ij} \cup b_{ij})]_{(m,n)} \quad and \quad A \cap B = [(a_{ij} \cap b_{ij})]_{(m,n)}.$$

The results of these binary operations are called the **union** *and* **intersection** *of A and B, respectively.*

Example 3. For the Boolean matrices with entries either 0 or 1 in which (see Example 1, Section 1.1)

$$0 \cap 0 = 0, \quad 0 \cap 1 = 1 \cap 0 = 0, \quad 1 \cap 1 = 1,$$

and

$$0 \cup 0 = 0, \quad 0 \cup 1 = 1 \cup 0 = 1, \quad 1 \cup 1 = 1,$$

$$\begin{bmatrix} 1 & 0 & 1 & 0 \\ 0 & 1 & 0 & 1 \end{bmatrix} \cup \begin{bmatrix} 1 & 0 & 0 & 1 \\ 1 & 1 & 0 & 0 \end{bmatrix} = \begin{bmatrix} 1 & 0 & 1 & 1 \\ 1 & 1 & 0 & 1 \end{bmatrix}.$$

Also,

$$\begin{bmatrix} 1 & 0 & 1 & 0 \\ 0 & 1 & 0 & 1 \end{bmatrix} \cap \begin{bmatrix} 1 & 0 & 0 & 1 \\ 1 & 1 & 0 & 0 \end{bmatrix} = \begin{bmatrix} 1 & 0 & 0 & 0 \\ 0 & 1 & 0 & 0 \end{bmatrix}.$$

It would be useful for the reader to show that the commutative and associative laws are valid for both operations $\cup$ and $\cap$ over the set of matrices whose entries are elements of a binary Boolean algebra. Both distributive laws

$$A \cap (B \cup C) = (A \cap B) \cup (A \cap C)$$

and

$$A \cup (B \cap C) = (A \cup B) \cap (A \cup C)$$

are also valid. The identity element for $\cup$ is the null matrix, that is,

$$A \cup 0 = 0 \cup A = A.$$

The identity element for $\cap$ is the matrix for which every entry is 1, that is,

$$A \cap U = U \cap A = A,$$

where $U = [1]_{(m,n)}$.

The next two binary operations to be defined make use of the fact that, in general, $AB \neq BA$. The symbol (j) is used to denote a binary operation known as **Jordan multiplication** (named for Marie Ennemond Camille Jordan, French, 1838–1922).

Definition 2.8. *If A and B are two n by n matrices, their* **Jordan** *product is*

$$A \textcircled{j} B = \frac{AB + BA}{2}.$$

Example 4.

$$\begin{bmatrix} 1 & 2 \\ 3 & 4 \end{bmatrix} \textcircled{j} \begin{bmatrix} 0 & 1 \\ 3 & 2 \end{bmatrix} = \frac{\begin{bmatrix} 1 & 2 \\ 3 & 4 \end{bmatrix}\begin{bmatrix} 0 & 1 \\ 3 & 2 \end{bmatrix} + \begin{bmatrix} 0 & 1 \\ 3 & 2 \end{bmatrix}\begin{bmatrix} 1 & 2 \\ 3 & 4 \end{bmatrix}}{2}$$

$$= \frac{1}{2}\begin{bmatrix} 6 & 5 \\ 12 & 11 \end{bmatrix} + \frac{1}{2}\begin{bmatrix} 3 & 4 \\ 9 & 14 \end{bmatrix} = \begin{bmatrix} \frac{9}{2} & \frac{9}{2} \\ \frac{21}{2} & \frac{25}{2} \end{bmatrix}.$$

It can be shown that the commutative property is valid for $\textcircled{j}$, but that the associative property is *not* valid. The reader should investigate both the distributive property for $\textcircled{j}$ with respect to $+$ and the cancellation property.

Algebras with operations that are nonassociative have become increasingly used in recent years. A wide range of applications occurs in such different fields as genetics and quantum mechanics.

The result of another nonassociative binary operation is known as a *matrix cross product;* this result is often called the *commutator* of two given matrices.

Definition 2.9. *If A and B are two n by n matrices, their matrix cross product is*

$$A \times B = AB - BA.$$

Example 5.

$$\begin{bmatrix} 1 & 2 \\ 3 & 4 \end{bmatrix} \times \begin{bmatrix} 0 & 1 \\ 3 & 2 \end{bmatrix} = \begin{bmatrix} 1 & 2 \\ 3 & 4 \end{bmatrix}\begin{bmatrix} 0 & 1 \\ 3 & 2 \end{bmatrix} - \begin{bmatrix} 0 & 1 \\ 3 & 2 \end{bmatrix}\begin{bmatrix} 1 & 2 \\ 3 & 4 \end{bmatrix}$$

$$= \begin{bmatrix} 6 & 5 \\ 12 & 11 \end{bmatrix} - \begin{bmatrix} 3 & 4 \\ 9 & 14 \end{bmatrix} = \begin{bmatrix} 3 & 1 \\ 3 & -3 \end{bmatrix}.$$

Some of the more important properties of this operation are summed up in the following theorem, the proof of which is left as an exercise for the reader.

Theorem 2.7. If A, B, C are n by n matrices and c is a scalar, then

$$A \times B = -(B \times A),$$
$$A \times A = 0,$$
$$A \times I = I \times A = 0,$$
$$A \times (B \times C) = B \times (A \times C) + C \times (B \times A),$$
$$A \times (B + C) = (A \times B) + (A \times C),$$
$$(A + B) \times C = (A \times C) + (B \times C),$$
$$c(A \times B) = (cA) \times B = A \times (cB).$$

APPLICATIONS

Example 6. If x and y represent elements of the set $\{0, 1\}$, the function

$$f(x, y) = x \cup (x \cap y)$$

is an example of a class of functions known as *binary Boolean functions.* This particular function $f(x, y)$ defined over the set of switches represents the circuit shown in Figure 2.3; this circuit is closed if x is closed (whether or not y is closed). Over the set of propositions, this function represents the statement

(proposition x) *or* {(proposition x) *and* (proposition y)} ;

this statement is true if proposition x is valid (whether or not proposition y is valid).

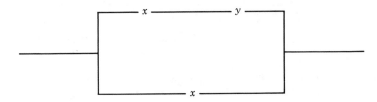

Fig. 2.3

The question arises, can this function $f(x, y) = x \cup (x \cap y)$ be represented by a matrix? If we let the variable x be represented by the matrix $\begin{bmatrix} 1 & 0 \\ 1 & 0 \end{bmatrix}$ and y be represented by the matrix $\begin{bmatrix} 1 & 1 \\ 0 & 0 \end{bmatrix}$, then we obtain results under the matrix operations $\cup$ and $\cap$ that are consistent with those obtained from Example 1 of Section 1.1 and Definition 2.5. For example,

$$f(x, y) = x \cup (x \cap y)$$

$$= \begin{bmatrix} 1 & 0 \\ 1 & 0 \end{bmatrix} \cup \left(\begin{bmatrix} 1 & 0 \\ 1 & 0 \end{bmatrix} \cap \begin{bmatrix} 1 & 1 \\ 0 & 0 \end{bmatrix} \right)$$

$$= \begin{bmatrix} 1 & 0 \\ 1 & 0 \end{bmatrix} \cup \begin{bmatrix} 1 & 0 \\ 0 & 0 \end{bmatrix} = \begin{bmatrix} 1 & 0 \\ 1 & 0 \end{bmatrix},$$

and we notice that the result is x; thus, we have a matrix representation for $f(x, y)$, and moreover a simplification has been accomplished by means of matrix operations.

Consider another binary Boolean function

$$f(x, y) = (x \cap y') \cup \{(x' \cup y) \cap x\}$$

which represents the arrangement of switches shown in Figure 2.4. Using the matrix representation for x and y and defining

$$x' = \begin{bmatrix} 0 & 1 \\ 0 & 1 \end{bmatrix} \qquad \text{and} \qquad y' = \begin{bmatrix} 0 & 0 \\ 1 & 1 \end{bmatrix}$$

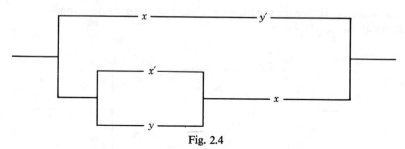

Fig. 2.4

(which are the opposites of x and y and satisfy the postulates of Definition 2.5), we have

$$f(x, y) = (x \cap y') \cup \{(x' \cup y) \cap x\}$$

$$= \left(\begin{bmatrix} 1 & 0 \\ 1 & 0 \end{bmatrix} \cap \begin{bmatrix} 0 & 0 \\ 1 & 1 \end{bmatrix} \right) \cup \left\{ \left(\begin{bmatrix} 0 & 1 \\ 0 & 1 \end{bmatrix} \cup \begin{bmatrix} 1 & 1 \\ 0 & 0 \end{bmatrix} \right) \cap \begin{bmatrix} 1 & 0 \\ 1 & 0 \end{bmatrix} \right\}$$

$$= \begin{bmatrix} 0 & 0 \\ 1 & 0 \end{bmatrix} \cup \left\{ \begin{bmatrix} 1 & 1 \\ 0 & 1 \end{bmatrix} \cap \begin{bmatrix} 1 & 0 \\ 1 & 0 \end{bmatrix} \right\}$$

$$= \begin{bmatrix} 0 & 0 \\ 1 & 0 \end{bmatrix} \cup \begin{bmatrix} 1 & 0 \\ 0 & 0 \end{bmatrix}$$

$$= \begin{bmatrix} 1 & 0 \\ 1 & 0 \end{bmatrix},$$

which is x. Hence, we have demonstrated that the single switch x is equivalent to the whole arrangement of switches shown in Figure 2.4. A manufacturer should install only switch x with the obvious advantage in savings in the cost, space requirements, and maintenance of the circuit. We have also demonstrated that a single proposition x is equivalent to a whole statement of propositions. Important simplifications or analyses of laws, contracts, and judicial decisions become possible.[1]

Notice that x and y are defined so that the following intersections

$$x \cap y = \begin{bmatrix} 1 & 0 \\ 0 & 0 \end{bmatrix}, \quad x \cap y' = \begin{bmatrix} 0 & 0 \\ 1 & 0 \end{bmatrix},$$

$$x' \cap y = \begin{bmatrix} 0 & 1 \\ 0 & 0 \end{bmatrix}, \quad x' \cap y' = \begin{bmatrix} 0 & 0 \\ 0 & 1 \end{bmatrix},$$

are represented by unique matrices each with exactly one nonzero entry. To define matrices representing x, y, and z for functions of three variables, matrices with eight entries are needed. The reader should verify that using the following definitions

$$x = \begin{bmatrix} 1 & 1 & 0 & 0 \\ 1 & 1 & 0 & 0 \end{bmatrix}, \quad y = \begin{bmatrix} 1 & 1 & 1 & 1 \\ 0 & 0 & 0 & 0 \end{bmatrix}, \quad z = \begin{bmatrix} 0 & 1 & 1 & 0 \\ 0 & 1 & 1 & 0 \end{bmatrix},$$

the following statements are valid,

$$x \cap (y \cap z) = \begin{bmatrix} 0 & 1 & 0 & 0 \\ 0 & 0 & 0 & 0 \end{bmatrix}, \quad \text{and} \quad x' \cap (y \cap z') = \begin{bmatrix} 0 & 0 & 0 & 1 \\ 0 & 0 & 0 & 0 \end{bmatrix}.$$

A more detailed discussion of Boolean matrices and their applications may be found elsewhere.[2]

Example 7. In quantum mechanics, the Pauli spin matrices (defined in Example 7 of Section 2.4) σ_x, σ_y, σ_z are very important. It can be verified that the following matrix cross products are valid:

$$\sigma_x \times \sigma_y = 2i\sigma_z,$$
$$\sigma_y \times \sigma_z = 2i\sigma_x,$$
$$\sigma_z \times \sigma_x = 2i\sigma_y.$$

[1] Fred Kort, "Simultaneous Equations and Boolean Algebra in the Analysis of Judicial Decisions" in *Jurimetrics*, Hans Baade, ed., New York, Basic Books, 1963.

[2] H. G. Flegg, *Boolean Algebra and Its Application* (New York, John Wiley & Sons, Inc., 1964).

EXERCISES

1. Consider the set of 2 by 2 matrices with entries that are elements of a binary Boolean algebra.

(*a*) What are the union and intersection of

$$\begin{bmatrix} 0 & 1 \\ 1 & 0 \end{bmatrix} \quad \text{and} \quad \begin{bmatrix} 0 & 0 \\ 1 & 1 \end{bmatrix}?$$

(*b*) What are the identity elements for $\cup$ and $\cap$?

(*c*) Prove that the commutative law for $\cup$ is valid.

2. Find the Jordan product and the cross product of

$$\begin{bmatrix} 3 & 1 \\ 2 & 4 \end{bmatrix} \quad \text{and} \quad \begin{bmatrix} 6 & 9 \\ 2 & 0 \end{bmatrix}.$$

3. Prove that the Jordan product is commutative but not associative.

4. Justify the binary Boolean algebra theorems

$$(x \cup y)' = x' \cap y' \quad \text{and} \quad (x \cap y)' = x' \cup y',$$

using Boolean matrices. (De Morgan's laws, named for Augustus De Morgan, English, 1806–1871.)

5. (*a*) Express the arrangement of switches shown in Figure 2.5 as a binary Boolean function.

(*b*) Simplify the function using Boolean matrices.

6. For a binary Boolean function of three variables $f(x, y, z)$, how many entries must a Boolean matrix representing x, y, or z contain?

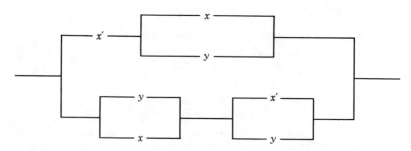

Fig. 2.5

7. Verify the equations of Example 7 of this section.

8. Prove the first three identities of Theorem 2.7.

9. Prove the last four identities of Theorem 2.7.

NEW VOCABULARY

§2.1　binary operation
§2.1　closed set under "∘"
§2.2　sum of two matrices
§2.2　conformable for addition
§§2.2, 2.4　commutative property
§§2.2, 2.4　associative property
§2.2　cancellation property
§2.3　multiplication of a matrix by a scalar
§2.3　subtraction of matrices
§2.3　identity matrix
§2.3　identity element
§2.3　scalar matrix

§2.4　product of two matrices
§2.4　conformable for multiplication
§2.4　distributive property
§2.4　stochastic matrix
§2.5　Boolean algebra
§2.5　Boolean matrix
§2.5　binary Boolean algebra
§2.5　union
§2.5　intersection
§2.5　Jordan product
§2.5　matrix cross product
§2.5　binary Boolean function

Unary Operations on Matrices

3.1 UNARY OPERATIONS

In Section 2.1 a binary operation was defined to be a correspondence by which any two elements are combined to produce a unique third entity. Suppose that, instead of performing some operation on two elements, we simply operate on one element of a set to produce a unique second element. Thus, "take the cube of" 2 produces 8, or "take the positive square root of" 16 produces 4. In trigonometry there are six very important operations that map one element into another entity, for example, "sine of" $\frac{1}{6}\pi$ produces $\frac{1}{2}$.

Definition 3.1. *A **unary operation** from a set S to a set R is a correspondence by which each element of S is assigned a unique element of R. If R is a subset of S, then we say that we have a **unary operation over S.***

As a further example, remember that earlier in this book we considered the "magnitude of" a vector, designated by $|\alpha|$ (Example 4, Section 1.3). Thus, $|(3, 0, 4)| = \sqrt{3^2 + 0^2 + 4^2} = 5$. Notice that the result of this unary operation does not belong to the same set as the recipient of the operation.

The purpose of this chapter is to discuss and define certain unary operations on matrices and to use these results.

3.2 THE TRANSPOSE OF A MATRIX

Let S consist of the set of all matrices. One useful unary operation over S consists of interchanging the rows and columns of a given matrix of this set.

Definition 3.2. *The **transpose** of a matrix A is a matrix that is formed by interchanging the rows and columns of A. The ith row of A*

47

becomes the ith column of the transpose of A. The transpose of A is denoted by A^T.

Notice that if A is an m by n matrix then the order of A^T is n by m.

Example 1.

$$A = \begin{bmatrix} 2 & 1 & 4 \\ 0 & 3 & 6 \end{bmatrix}, \qquad A^T = \begin{bmatrix} 2 & 0 \\ 1 & 3 \\ 4 & 6 \end{bmatrix}.$$

Next we should ask: What are the properties of this new operation $(\ )^T$? In this text we list four properties. Proofs of the first three theorems are left as exercises for the reader.

Theorem 3.1. $(A^T)^T = A$.

Theorem 3.2.   *If A is conformable to B for addition then*

$$(A + B)^T = A^T + B^T.$$

Theorem 3.3. $(cA)^T = cA^T$,
where c is a scalar.

Theorem 3.4.   *If A is conformable to B for multiplication then*

$$(AB)^T = B^T A^T.$$

Proof when A and B are real.   We prove that the two matrices are equal by showing that the corresponding entries are equal. Let $C = AB$, where A has order m by k and B has order k by n.

STATEMENT	REASON
(1) The (ij)th entry of C^T (that is, of $(AB)^T$) $= (ji)$th entry of C	(1) Definition 3.2.
(2) $= (j$th row of $A)\cdot(i$th column of $B)$	(2) Matrix multiplication.
(3) $= (a_{j1}, a_{j2}, \ldots, a_{jk})\cdot(b_{1i}, b_{2i}, \ldots, b_{ki})$	(3) Scalar product of two vectors.
(4) $= (b_{1i}, b_{2i}, \ldots, b_{ki})\cdot(a_{j1}, a_{j2}, \ldots, a_{jk})$	(4) Scalar product of two vectors is commutative.

(5) $= (i$th row of $B^T) \cdot (j$th column of $A^T)$　　　(5) Definition 3.2.

(6) $= (ij)$th entry of $B^T A^T$.　　　(6) Matrix multiplication.

(7) $(AB)^T$ and $B^T A^T$ have the same order.　　　(7) Why?

(8) $(AB)^T = B^T A^T$.　　　(8) Definition of equality of matrices.

Example 2. Let

$$A = \begin{bmatrix} 1 & 0 \\ 2 & 1 \end{bmatrix} \quad \text{and} \quad B = \begin{bmatrix} 0 & 1 & 2 \\ 1 & 1 & 3 \end{bmatrix}.$$

$$(AB)^T = \begin{bmatrix} 0 & 1 & 2 \\ 1 & 3 & 7 \end{bmatrix}^T = \begin{bmatrix} 0 & 1 \\ 1 & 3 \\ 2 & 7 \end{bmatrix};$$

$$B^T A^T = \begin{bmatrix} 0 & 1 \\ 1 & 1 \\ 2 & 3 \end{bmatrix} \begin{bmatrix} 1 & 2 \\ 0 & 1 \end{bmatrix} = \begin{bmatrix} 0 & 1 \\ 1 & 3 \\ 2 & 7 \end{bmatrix}.$$

We are now in a position to define several special types of matrices that are useful in both applied and theoretical matrix mathematics.

Definition 3.3.　*A matrix A is **symmetric** if $A = A^T$.*

Definition 3.4.　*A matrix A is **skew-symmetric** if $A = -A^T$.*

Definition 3.5.　*A matrix with entries that are complex conjugates of the corresponding entries of A is the **conjugate** of A and is denoted by $\bar{A} = [\bar{a}_{ij}]_{(m,n)}$.*

Definition 3.6.　*A matrix A is **Hermitian**[1] if $A = \bar{A}^T$.*

Definition 3.7.　*A matrix A is **skew-Hermitian** if $A = -\bar{A}^T$.*

Example 3.

$A = \begin{bmatrix} 4 & 1 & 6 \\ 1 & 2 & 0 \\ 6 & 0 & 3 \end{bmatrix}$ is symmetric because $A = A^T$.

---

[1] Named for Charles Hermite, French, 1822–1901.

$$B = \begin{bmatrix} 0 & 1 & 3 \\ -1 & 0 & -2 \\ -3 & 2 & 0 \end{bmatrix} \text{ is skew-symmetric because } B = -B^{\mathrm{T}}.$$

$$C = \begin{bmatrix} 4 & 3+i \\ 3-i & 2 \end{bmatrix}, \quad \overline{C} = \begin{bmatrix} 4 & 3-i \\ 3+i & 2 \end{bmatrix}, \quad \overline{C}^{\mathrm{T}} = \begin{bmatrix} 4 & 3+i \\ 3-i & 2 \end{bmatrix}.$$

Therefore C is Hermitian because $C = \overline{C}^{\mathrm{T}}$.

$$D = \begin{bmatrix} 0 & -4+i \\ 4+i & 0 \end{bmatrix}, \quad \overline{D} = \begin{bmatrix} 0 & -4-i \\ 4-i & 0 \end{bmatrix},$$

$$\overline{D}^{\mathrm{T}} = \begin{bmatrix} 0 & -(-4+i) \\ -(4+i) & 0 \end{bmatrix}.$$

Therefore D is skew-Hermitian because $D = -\overline{D}^{\mathrm{T}}$.

APPLICATIONS

★ **Example 4.** In many engineering and social science problems, quadratic expressions of a form similar to

$$f(x, y, z) = x^2 + 4xy + 2y^2 + 6yz + 4z^2 + xz$$

appear. It is possible to express $f(x, y, z)$ using matrix notation as follows:

$$[f(x, y, z)] = X^{\mathrm{T}}AX,$$

where

$$A = \begin{bmatrix} 1 & 2 & \frac{1}{2} \\ 2 & 2 & 3 \\ \frac{1}{2} & 3 & 4 \end{bmatrix} \quad \text{and} \quad X = \begin{bmatrix} x \\ y \\ z \end{bmatrix}.$$

The reader should verify this multiplication. Notice that the unary operation ( )^{T} was used and that A is a symmetric matrix which is constructed by letting the entries along the main diagonal represent the coefficients of the squared terms of $f(x, y, z)$, and the other entries are one-half of the coefficients of the corresponding product terms. For those who are faced with problems of this general type it is useful to investigate the properties of symmetric matrices.

Example 5. Consider the number of roads connecting the four cities shown in the diagram (Figure 3.1). (Of course, many other interpretations besides roads may be given to the diagram, as for instance, communication

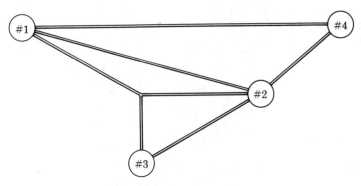

Fig. 3.1

lines or mutual influence among people or nations.) This information can be represented by a matrix A, where

$a_{ij} = a_{ji} =$ (NUMBER OF ROADS CONNECTING THE ith
                  AND jth CITIES WITHOUT PASSING THROUGH ANOTHER CITY),

$$A = \begin{array}{cc} \begin{array}{cccc} \#1 & \#2 & \#3 & \#4 \end{array} & \\ \begin{bmatrix} 0 & 2 & 1 & 1 \\ 2 & 0 & 2 & 1 \\ 1 & 2 & 0 & 0 \\ 1 & 1 & 0 & 0 \end{bmatrix} & \begin{array}{c} \#1 \\ \#2 \\ \#3 \\ \#4 \end{array} \end{array}.$$

A is a symmetric matrix. It can be shown by a method similar to that of Example 6 of Section 2.4 that A^2 represents the number of ways to travel between any two cities by passing through exactly one city; opposite directions on the same road are considered different. A^3 represents the number of ways to travel between any two cities by passing through exactly two cities. It is left as an exercise to show that if A is symmetric then A^n, where n is a positive integer, is symmetric. Also $A + A^2$ (which represents the total number of ways to travel between two cities with at most one intermediate city) is symmetric.

Example 6. Any square matrix A can be expressed as the sum of a symmetric matrix S and a skew-symmetric matrix K. If A is any square matrix, then S and K can be constructed as follows:

$$S = \frac{A + A^T}{2}, \quad K = \frac{A - A^T}{2}, \quad \text{and} \quad A = S + K.$$

For example, if $A = \begin{bmatrix} 1 & 0 \\ 2 & 4 \end{bmatrix}$, then $A^T = \begin{bmatrix} 1 & 2 \\ 0 & 4 \end{bmatrix}$,

$$S = \frac{\begin{bmatrix} 1 & 0 \\ 2 & 4 \end{bmatrix} + \begin{bmatrix} 1 & 2 \\ 0 & 4 \end{bmatrix}}{2} = \begin{bmatrix} 1 & 1 \\ 1 & 4 \end{bmatrix},$$

$$K = \frac{\begin{bmatrix} 1 & 0 \\ 2 & 4 \end{bmatrix} - \begin{bmatrix} 1 & 2 \\ 0 & 4 \end{bmatrix}}{2} = \begin{bmatrix} 0 & -1 \\ 1 & 0 \end{bmatrix}.$$

Thus, $\qquad A = S + K = \begin{bmatrix} 1 & 1 \\ 1 & 4 \end{bmatrix} + \begin{bmatrix} 0 & -1 \\ 1 & 0 \end{bmatrix} = \begin{bmatrix} 1 & 0 \\ 2 & 4 \end{bmatrix}.$

Example 7. In the theory of relativity, the matrix

$$\begin{bmatrix} 1 & 0 & 0 & 0 \\ 0 & 1 & 0 & 0 \\ 0 & 0 & \gamma & \dfrac{iv\gamma}{c} \\ 0 & 0 & \dfrac{-iv\gamma}{c} & \gamma \end{bmatrix}$$

is a transformation matrix from a stationary system to a system moving with a velocity v with respect to the stationary system. Here c represents the velocity of light, and $\gamma = 1/\sqrt{1 - (v^2/c^2)}$. Notice that the matrix is Hermitian.

Example 8. The Pauli spin matrices σ_x, σ_y, and σ_z (defined in Example 7 of Section 2.4) are Hermitian. It can be shown that the products of any two of the Pauli spin matrices are skew-Hermitian.

EXERCISES

1. Determine which of the following are symmetric or skew-symmetric. If any one is neither, state why it is neither.

(a) $\begin{bmatrix} 1 & 2 & 5 \\ 2 & 2 & -1 \\ 5 & -1 & 3 \end{bmatrix}$;

(b) $\begin{bmatrix} 2 & 3 \\ 3 & 4 \\ 0 & 0 \end{bmatrix}$;

(c) $\begin{bmatrix} 0 & 0 & 2 \\ 0 & 0 & -1 \\ -2 & 1 & 0 \end{bmatrix}$;

(d) $\begin{bmatrix} 1 & -2 & 3 \\ 2 & 1 & 4 \\ -3 & -4 & 1 \end{bmatrix}$.

2. When is A conformable to A^T for addition? For multiplication?

3. Using $A = \begin{bmatrix} 3 & 2 \\ 4 & 0 \end{bmatrix}$ and $B = \begin{bmatrix} 0 & 1 & 2 \\ 1 & 2 & 0 \end{bmatrix}$ verify:

   (a) Theorem 3.1;   (b) Theorem 3.4.

4. Does $(AB)^T = A^T B^T$? Why?

5. Does $(A - B)^T = A^T - B^T$? If not, give a counterexample. If the relation is true, prove it using Theorems 3.2 and 3.3.

6. Simplify: $(A^T B^T + 3C)^T$. Give the reason for each step.

7. Which of the following matrices are Hermitian? Which are skew-Hermitian? Which are neither and why?

$$(a) \begin{bmatrix} i & 2-i \\ 0 & 3i \\ 2 & 4 \end{bmatrix} ; \quad (b) \begin{bmatrix} 2 & 2+i \\ i & 2i \end{bmatrix} ; \quad (c) \begin{bmatrix} 0 & 3+i \\ 3-i & 2 \end{bmatrix} ;$$

$$(d) \begin{bmatrix} 2 & 3 \\ 3 & 1 \end{bmatrix} ; \quad (e) \begin{bmatrix} 0 & i & 2i \\ i & 0 & -4 \\ 2i & 4 & 0 \end{bmatrix} .$$

8. Prove that each entry on the main diagonal of a skew-symmetric matrix with complex entries must be 0.

9. Prove that if A is a square matrix, then $A + A^T$ is symmetric.

10. Prove that if A is symmetric, then A^2 is symmetric.

11. Prove that if A is skew-symmetric, then A^2 is symmetric.

12. Prove that if A and B are symmetric and if $AB = BA$, then AB is symmetric.

13. Prove that if A is a skew-symmetric matrix, then $AA^T = A^T A$.

14. Show that the product of two symmetric matrices of the same order need not be symmetric. (*Hint:* Use an example.)

15. (a) Prove Theorem 3.1.
    (b) Prove Theorem 3.2.
    (c) Prove Theorem 3.3.

16. Express the following matrix as the sum of a skew-symmetric matrix and a symmetric matrix:

$$\begin{bmatrix} 3 & 3 & -1 \\ 0 & 3 & -2 \\ -1 & 2 & 2 \end{bmatrix}.$$

17. Express the quadratic form

$$f(x, y, z) = x^2 - 4xy + y^2 + 6yz + z^2$$

in (symmetric) matrix form.

18. Consider a small telephone system between 3 towns in which there are 5 lines between town #2 and town #3; 3 lines between town #1 and town #3, and 4 lines between town #1 and town #2. Use matrices to determine the total number of different communication channels between each pair of the towns; a channel between two towns may or may not pass through a third town.

19. Verify the assertions of Example 8.

3.3 THE TRACE OF A MATRIX

A unary operation which has some interesting properties and which is useful in matrix theory is the *trace* of a matrix.

Definition 3.8. *The trace of a square matrix A is the sum of the entries on the main diagonal and is denoted as* tr *A.*

Example 1.

$$\text{tr} \begin{bmatrix} a_{11} & a_{12} & a_{13} \\ a_{21} & a_{22} & a_{23} \\ a_{31} & a_{32} & a_{33} \end{bmatrix} = a_{11} + a_{22} + a_{33}.$$

$$\text{tr} \begin{bmatrix} 2 & 0 & 1 \\ 4 & 3 & 6 \\ 9 & 8 & 5 \end{bmatrix} = 2 + 3 + 5 = 10.$$

Notice that the result of this unary operation on a square matrix with scalar entries is a scalar. Some properties of the operation tr ( ) are listed below; the proofs of the first three theorems are left as exercises.

Theorem 3.5. *If A is a square matrix, then*

$$\text{tr} \, A = \text{tr} \, A^{\text{T}}.$$

Theorem 3.6. *If A and B are n by n matrices, then*

$$\operatorname{tr}(A + B) = \operatorname{tr} A + \operatorname{tr} B.$$

Theorem 3.7. *If A is a n by n matrix and c is a scalar, then*

$$\operatorname{tr}(cA) = c \operatorname{tr} A.$$

Theorem 3.8. *If A and B are n by n matrices, then*

$$\operatorname{tr} AB = \operatorname{tr} BA.$$

Proof. See Theorem A.4, page 156, in Section A.6 following a discussion of the Σ notation in Sections A.1 and A.2.

Example 2. Let

$$A = \begin{bmatrix} 4 & 0 & 6 \\ 5 & 2 & 1 \\ 7 & 8 & 3 \end{bmatrix}, \qquad B = \begin{bmatrix} 1 & 0 & 1 \\ 9 & 1 & 2 \\ 0 & 4 & 1 \end{bmatrix}.$$

(a) $\operatorname{tr} A = 4 + 2 + 3 = 9;$ $\operatorname{tr} A^{\mathrm{T}} = 4 + 2 + 3 = 9.$

(b) $\operatorname{tr}(A + B) = 5 + 3 + 4 = 12;$ $\operatorname{tr} A + \operatorname{tr} B = 9 + 3 = 12.$

(c) $\operatorname{tr}(AB) = \operatorname{tr} \begin{bmatrix} 4 & 24 & 10 \\ 23 & 6 & 10 \\ 79 & 20 & 26 \end{bmatrix} = 36;$ $\operatorname{tr}(BA) = \operatorname{tr} \begin{bmatrix} 11 & 8 & 9 \\ 55 & 18 & 61 \\ 27 & 16 & 7 \end{bmatrix} = 36.$

APPLICATIONS

Example 3. One simple application of the trace of a matrix can be obtained from Example 6 of Section 2.4. The trace of the sum $A + A^2 + \cdots + A^{n+1}$ represents the total number of communication channels by which a message may return to its origin after at most n relays.

Example 4. Suppose that a certain transportation industry maintains a regular service between n stations. Let the entry a_{ij} of matrix A represent the number of passengers transported from station i to station j, and let the entry b_{ij} of matrix B represent the cost of a ticket from station i to station j. The trace of matrix AB^{T} (or AB if B is symmetric) represents the total income of the industry from this service.

Other applications of the trace of a matrix that are important in economics and engineering are given in Section 8.2.

EXERCISES

1. Find the trace of $\begin{bmatrix} 3 & 0 & 1 \\ 4 & 2 & 2 \\ 0 & 9 & 1 \end{bmatrix}$.

2. Verify the theorems of this section for the matrices

$$A = \begin{bmatrix} 2 & 1 \\ 4 & 6 \end{bmatrix}, \qquad B = \begin{bmatrix} 2 & 0 \\ 3 & 4 \end{bmatrix}.$$

3. Prove: (a) Theorem 3.5; (b) Theorem 3.6; (c) Theorem 3.7.

4. Consider three cities that are connected by roads as shown in Figure 3.2. Let $a_{ij} = a_{ji}$ represent the number of routes between cities i and j without passing through another city (see Example 5 of Section 3.2). Use the diagram to verify that the tr A^2 represents the total number of possible round trips originating from all cities and passing through exactly one other city.

5. In the preceding exercise explain the meaning of tr $(A^3 + A^2 + A)$.

6. In Example 4 of this section verify the last sentence for $n = 3$ or 3 stations.

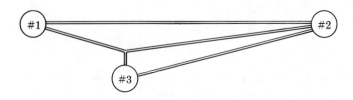

Fig. 3.2

3.4 THE DETERMINANT OF A MATRIX

In this section we investigate another unary operation on a *square matrix* that produces a scalar. The scalar resulting from this operation is called the **determinant** of the matrix and is denoted by either det A or $|A|$. The determinant of an n by n matrix is an **nth-order determinant**.

Definition 3.9. *If the ith row and jth column of an n by n matrix A are deleted ($n \geq 2$), the determinant of the resulting submatrix is called* the **minor** of a_{ij} and is denoted by M_{ij}. The quantity $A_{ij} = (-1)^{i+j}M_{ij}$ is called the **cofactor** of a_{ij}.

Example 1. For

$$A = \begin{bmatrix} a_{11} & a_{12} & a_{13} \\ a_{21} & a_{22} & a_{23} \\ a_{31} & a_{32} & a_{33} \end{bmatrix},$$

the minor of a_{23} is

$$M_{23} = \det \begin{bmatrix} a_{11} & a_{12} \\ a_{31} & a_{32} \end{bmatrix}$$

and the cofactor of a_{23} is

$$A_{23} = (-1)^{2+3}M_{23} = (-1) \det \begin{bmatrix} a_{11} & a_{12} \\ a_{31} & a_{32} \end{bmatrix}.$$

Definition 3.10.   *If a matrix contains a single entry, then the determinant of the matrix is equal to that entry. If A is a square matrix of order n, $(n \geq 2)$, where i and j are fixed integers, and $1 \leq i \leq n$, $1 \leq j \leq n$, then its **determinant** is equal to either*

$$|A| = \det A = a_{i1}A_{i1} + a_{i2}A_{i2} + \cdots + a_{in}A_{in},$$

or

$$|A| = \det A = a_{1j}A_{1j} + a_{2j}A_{2j} + \cdots + a_{nj}A_{nj}.$$

The validity of this definition depends on the following theorem.

Theorem 3.9.   *The value of $\det A$ is the same no matter which row i or column j is chosen.*

A proof of this theorem as well as proofs of the later theorems (except Theorem 3.14[2]) of this chapter may be found in many college algebra texts.[3]

If $A = \begin{bmatrix} a_{11} & a_{12} \\ a_{21} & a_{22} \end{bmatrix}$, then $\det A$ can be written as $|A| = \begin{vmatrix} a_{11} & a_{12} \\ a_{21} & a_{22} \end{vmatrix}$. In general the determinant of any square matrix can be expressed as a square array of entries enclosed by vertical lines.

---

[2] A proof of Theorem 3.14 is found in S. Perlis, *Theory of Matrices*, Reading, Massachusetts, Addison-Wesley Publishing Company, Inc., 1952, pp. 79–80. A knowledge of elementary matrices, which are to be studied in Chapter 5, is needed to read this proof.

[3] R. W. Brink, *College Algebra* 2nd ed., New York, Appleton-Century-Crofts, 1951.

Example 2. From Definition 3.10,

$$\det \begin{bmatrix} a_{11} & a_{12} \\ a_{21} & a_{22} \end{bmatrix} = \begin{vmatrix} a_{11} & a_{12} \\ a_{21} & a_{22} \end{vmatrix}$$

$$= a_{11}(-1)^{1+1}|a_{22}| + a_{12}(-1)^{1+2}|a_{21}|$$

$$= a_{11}a_{22} - a_{12}a_{21}.$$

In the first (second) expression given in Definition 3.10 we say that det A has been **expanded** about the ith row (jth column), or we may call this expression the **expansion** of det A about the ith row (jth column).

Example 3. The expansion of

$$\det \begin{bmatrix} 2 & 1 & 4 \\ 1 & 0 & 6 \\ 2 & 3 & 0 \end{bmatrix} \quad \text{or} \quad \begin{vmatrix} 2 & 1 & 4 \\ 1 & 0 & 6 \\ 2 & 3 & 0 \end{vmatrix}$$

about the 3rd row is

$$2(-1)^{3+1}\det \begin{bmatrix} 1 & 4 \\ 0 & 6 \end{bmatrix} + 3(-1)^{3+2}\det \begin{bmatrix} 2 & 4 \\ 1 & 6 \end{bmatrix} + 0(-1)^{3+3}\det \begin{bmatrix} 2 & 1 \\ 1 & 0 \end{bmatrix}$$

$$= 2(1 \cdot 6 - 4 \cdot 0) + 3(-1)(2 \cdot 6 - 4 \cdot 1) + 0 = -12.$$

Example 4. Evaluate

$$|A| = \begin{vmatrix} 3 & 4 & 6 & 1 \\ 0 & 1 & 0 & 3 \\ 0 & 1 & 0 & 4 \\ 1 & -2 & 1 & 3 \end{vmatrix}.$$

If we expand $|A|$ about the 3rd column, we obtain

$$6(-1)^{1+3}\begin{vmatrix} 0 & 1 & 3 \\ 0 & 1 & 4 \\ 1 & -2 & 3 \end{vmatrix} + 0 + 0 + (1)(-1)^{4+3}\begin{vmatrix} 3 & 4 & 1 \\ 0 & 1 & 3 \\ 0 & 1 & 4 \end{vmatrix}$$

$$= 6\begin{vmatrix} 0 & 1 & 3 \\ 0 & 1 & 4 \\ 1 & -2 & 3 \end{vmatrix} - \begin{vmatrix} 3 & 4 & 1 \\ 0 & 1 & 3 \\ 0 & 1 & 4 \end{vmatrix}.$$

Both of these minors should be expanded about the first column.

$$|A| = 6\left(0 + 0 + (1)(-1)^{3+1}\begin{vmatrix} 1 & 3 \\ 1 & 4 \end{vmatrix}\right) - \left((3)(-1)^{1+1}\begin{vmatrix} 1 & 3 \\ 1 & 4 \end{vmatrix} + 0 + 0\right)$$

$$= 6(1 \cdot 4 - 3 \cdot 1) - 3(1 \cdot 4 - 3 \cdot 1) = 3.$$

The properties of determinants which prove useful to us later in this book are stated below.

Theorem 3.10.   *If a matrix B is formed from a matrix A by the interchange of two parallel lines (that is, rows or columns), then $|A| = -|B|$.*

Corollary.   *If two parallel lines (rows or columns) of a matrix are identical, then the determinant of the matrix is equal to zero.*

Theorem 3.11.   *The determinants of a matrix and its transpose are equal; that is, $|A| = |A^{T}|$.*

Theorem 3.12.   *The sum of the products of the entries of one line of A and the cofactors of the corresponding entries of a different parallel line of A is zero; that is,*

$$a_{i1}A_{k1} + a_{i2}A_{k2} + \cdots + a_{in}A_{kn} = 0, \qquad \text{if } i \neq k,$$

$$a_{1j}A_{1k} + a_{2j}A_{2k} + \cdots + a_{nj}A_{nk} = 0, \qquad \text{if } j \neq k.$$

Example 5.   If $|A|$ is expanded about the first column,

$$|A| = \begin{vmatrix} 2 & 4 & 1 \\ 3 & 4 & 6 \\ 4 & 0 & 1 \end{vmatrix} = 2\begin{vmatrix} 4 & 6 \\ 0 & 1 \end{vmatrix} - 3\begin{vmatrix} 4 & 1 \\ 0 & 1 \end{vmatrix} + 4\begin{vmatrix} 4 & 1 \\ 4 & 6 \end{vmatrix}.$$

However, if we replace the elements of the first column in the expansion by the elements of any other column, say the second column, the value is zero; that is,

$$4\begin{vmatrix} 4 & 6 \\ 0 & 1 \end{vmatrix} - 4\begin{vmatrix} 4 & 1 \\ 0 & 1 \end{vmatrix} + 0\begin{vmatrix} 4 & 1 \\ 4 & 6 \end{vmatrix} = 0.$$

Theorem 3.13.   *The value of a determinant $|A|$ is multiplied by a scalar c whenever every entry of a row (or column) of A is multiplied by c.*

Example 6.

(a) $c|B| = c\begin{vmatrix} b_{11} & b_{12} & b_{13} \\ b_{21} & b_{22} & b_{23} \\ b_{31} & b_{32} & b_{33} \end{vmatrix} = \begin{vmatrix} cb_{11} & cb_{12} & cb_{13} \\ b_{21} & b_{22} & b_{23} \\ b_{31} & b_{32} & b_{33} \end{vmatrix} = \begin{vmatrix} b_{11} & cb_{12} & b_{13} \\ b_{21} & cb_{22} & b_{23} \\ b_{31} & cb_{32} & b_{33} \end{vmatrix}.$

(b)   If $A = \begin{bmatrix} 2 & 1 \\ 2 & 3 \end{bmatrix}$, then $2|A| = \begin{vmatrix} 2 & 1 \\ 4 & 6 \end{vmatrix}.$

Theorem 3.14. *The determinant of the product of two square matrices of the same order is equal to the product of the determinants of the two matrices; that is, $|AB| = |A||B|$.*

Theorem 3.15. *If a matrix B is obtained from a matrix A by adding to each entry of a line of A a constant multiple of the corresponding entry of a parallel line, then $|B| = |A|$.*

Example 7. Let

$$A = \begin{bmatrix} 1 & 0 & 1 \\ 0 & 4 & 6 \\ -2 & 2 & 1 \end{bmatrix}.$$

If we multiply each entry of the first row of A by 2 and add the results to the corresponding entries of the third row (abbreviated $2R_1 + R_3$), there results a matrix whose determinant is equal to det A. That is,

$$\begin{vmatrix} 1 & 0 & 1 \\ 0 & 4 & 6 \\ -2 & 2 & 1 \end{vmatrix}_{2R_1+R_3} = \begin{vmatrix} 1 & 0 & 1 \\ 0 & 4 & 6 \\ 0 & 2 & 3 \end{vmatrix}.$$

Notice that the only row that changed was the third row — the *recipient* of the operation. The student should also observe that an operation such as $2R_1 + 3R_3$ is not covered by Theorem 3.15; the coefficient of the recipient line in the notation $kR_j + R_i$, must be 1.

Theorem 3.15 proves to be of great help in the evaluation of higher-order determinants. For instance, in Example 7, the original third-order determinant has essentially been reduced to a constant times a second-order determinant if it is expanded about its first column.

Example 8. Evaluate

$$\begin{vmatrix} 198 & 0 & 99 & 99 \\ 1 & 1 & -2 & 0 \\ 1 & 2 & 1 & 2 \\ 1 & -3 & 6 & 1 \end{vmatrix}.$$

By Theorems 3.13 and 3.15, we have

$$99\begin{vmatrix} 2 & 0 & 1 & 1 \\ 1 & 1 & -2 & 0 \\ 1 & 2 & 1 & 2 \\ 1 & -3 & 6 & 1 \end{vmatrix}_{-2R_2+R_3} = 99\begin{vmatrix} 2 & 0 & 1 & 1 \\ 1 & 1 & -2 & 0 \\ -1 & 0 & 5 & 2 \\ 1 & -3 & 6 & 1 \end{vmatrix}$$

$$\underset{3R_2+R_4}{=} \quad 99 \begin{vmatrix} 2 & 0 & 1 & 1 \\ 1 & 1 & -2 & 0 \\ -1 & 0 & 5 & 2 \\ 4 & 0 & 0 & 1 \end{vmatrix} \underset{(\text{Expand on } C_2)}{=} \quad 99(-1)^{2+2}(1) \begin{vmatrix} 2 & 1 & 1 \\ -1 & 5 & 2 \\ 4 & 0 & 1 \end{vmatrix}$$

$$\underset{-4C_3+C_1}{=} \quad 99 \begin{vmatrix} -2 & 1 & 1 \\ -9 & 5 & 2 \\ 0 & 0 & 1 \end{vmatrix} \underset{(\text{Expand on } R_3)}{=} \quad 99(-1)^{3+3}(1) \begin{vmatrix} -2 & 1 \\ -9 & 5 \end{vmatrix} = -99.$$

Definition 3.11. *A square matrix is said to be* **triangular** *if all the entries either above or below the main diagonal are zero. If all the entries both above and below the main diagonal are zero, the matrix is diagonal.*

An efficient way of calculating the determinant of a matrix with a computer is, using Theorem 3.15, to triangularize the matrix, and then the determinant of the resulting matrix is equal to the product of the entries along its main diagonal, which is relatively easy to compute.

Example 9.

$$\begin{vmatrix} 2 & 2 & 3 \\ 4 & 5 & 7 \\ 0 & 6 & 9 \end{vmatrix} \underset{-2R_1+R_2}{=} \begin{vmatrix} 2 & 2 & 3 \\ 0 & 1 & 1 \\ 0 & 6 & 9 \end{vmatrix} \underset{-6R_2+R_3}{=} \begin{vmatrix} 2 & 2 & 3 \\ 0 & 1 & 1 \\ 0 & 0 & 3 \end{vmatrix}$$

$$= 2 \begin{vmatrix} 1 & 1 \\ 0 & 3 \end{vmatrix} = 2 \cdot 1 \cdot 3 = 6.$$

APPLICATIONS

The origin of determinants is closely associated with techniques for solving systems of linear equations. There is evidence that before the time of Christ, the Chinese used bamboo rods to develop methods of solving simultaneous equations. Their methods seem to have been similar to what is now called the expansion of a determinant. Determinants did not begin to take definite form, however, until about 1683 in Japan and about 1693 in Germany. Seki Kowa (Japanese mathematician) and Leibniz were responsible. Their work was amplified in 1750 by Gabriel Cramer (Swiss, 1704–1752). The famous theorem which bears his name follows.

Theorem 3.16.   *Cramer's Rule.* *For the system*

$$\begin{cases} a_{11}x_1 + a_{12}x_2 + \cdots + a_{1n}x_n = b_1, \\ \qquad\qquad\qquad \cdot \\ \qquad\qquad\qquad \cdot \\ \qquad\qquad\qquad \cdot \\ a_{n1}x_1 + a_{n2}x_2 + \cdots + a_{nn}x_n = b_n, \end{cases} \quad \text{or} \quad AX = B,$$

let ^{j}A denote the matrix obtained from A by replacing the jth column of A by the vector B. If $\det A \neq 0$, *then the system $AX = B$ has exactly one solution; this solution is*

$$x_j = \frac{\det(^{j}A)}{\det A}, \qquad j = 1, 2, \ldots, n.$$

Proof.   See Theorem A.5, page 156, in Section A.7.

Example 10.   Solve by Cramer's rule the following system:

$$\begin{cases} 2x_1 + x_2 + x_3 = 0, \\ x_1 - x_2 + 5x_3 = 0, \\ x_2 - x_3 = 4. \end{cases}$$

$$x_1 = \frac{\begin{vmatrix} 0 & 1 & 1 \\ 0 & -1 & 5 \\ 4 & 1 & -1 \end{vmatrix}}{\begin{vmatrix} 2 & 1 & 1 \\ 1 & -1 & 5 \\ 0 & 1 & -1 \end{vmatrix}} = \frac{4\begin{vmatrix} 1 & 1 \\ -1 & 5 \end{vmatrix}}{-6} = \frac{24}{-6} = -4.$$

$$x_2 = \frac{\begin{vmatrix} 2 & 0 & 1 \\ 1 & 0 & 5 \\ 0 & 4 & -1 \end{vmatrix}}{\det A} = \frac{-4\begin{vmatrix} 2 & 1 \\ 1 & 5 \end{vmatrix}}{-6} = \frac{-36}{-6} = 6.$$

$$x_3 = \frac{\begin{vmatrix} 2 & 1 & 0 \\ 1 & -1 & 0 \\ 0 & 1 & 4 \end{vmatrix}}{\det A} = \frac{4\begin{vmatrix} 2 & 1 \\ 1 & -1 \end{vmatrix}}{-6} = \frac{-12}{-6} = 2.$$

Although the application demonstrated in Example 10 is interesting and has certain theoretical ramifications, it is not the best method for solving large systems of linear equations. We consider better methods in Chapter 5.

Example 11. One very important elementary application of determinants is their usefulness in simplifying notation. Consider the calculation of the area of the triangle shown in Figure 3.3. To do this, one calculates the areas of the three trapezoids shown in Figure 3.3 and combining these results one can verify that the area of the triangle is

$$\text{Area} = \frac{y_1 + y_3}{2}(x_3 - x_1) + \frac{y_3 + y_2}{2}(x_2 - x_3) - \frac{y_1 + y_2}{2}(x_2 - x_1).$$

It can be shown (Exercise 19) that the expansion of this formula is the same as the expansion of the determinant

$$\frac{1}{2}\begin{vmatrix} x_1 & y_1 & 1 \\ x_2 & y_2 & 1 \\ x_3 & y_3 & 1 \end{vmatrix}.$$

This is just one of many examples of the use of determinant notation to express complicated formulas.

Several other applications of determinants are illustrated later in this book, namely, in the calculation of the inverse of a given matrix, in a definition of rank which is useful in a general discussion of systems of linear equations, and, finally, in the definition of the characteristic equation of a matrix. There are many other applications of determi-

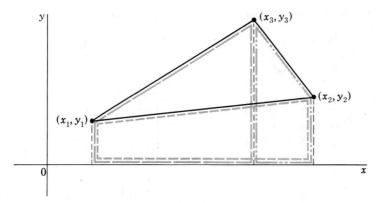

Fig. 3.3

nants which are beyond the scope of this book such as in the study of tensor algebra, modern geometry, advanced calculus, vector analysis, statistics, and matrix theory.

EXERCISES

1. Given

$$A = \begin{bmatrix} 3 & 1 & -4 \\ 6 & 9 & -2 \\ -1 & 2 & 1 \end{bmatrix}.$$

Without the aid of Theorem 3.15:
   (a) Expand $|A|$ about the first column.
   (b) Expand $|A|$ about the third row.
   (c) Expand $|A|$ about the third column.
   (d) What is the cofactor of the entry in the third row and second column?
   (e) What is the minor of the entry in the first row and second column?

2. Change the form but not the value of

$$|A| = \begin{vmatrix} -1 & 1 & 2 & 0 \\ -2 & 1 & 3 & 1 \\ 1 & 0 & 2 & -1 \\ 2 & 1 & -1 & 2 \end{vmatrix}$$

so that zeros occur everywhere in the first column except in the third row. (*Hint:* Use Theorem 3.15.)

3. Evaluate $|A|$ in Exercise 2.

4. With the aid of Theorem 3.15 evaluate each determinant:

$$(a) \begin{vmatrix} 3 & -1 & 3 \\ 2 & 5 & -3 \\ 5 & 4 & -1 \end{vmatrix}; \qquad (b) \begin{vmatrix} 1 & 1 & 1 & 1 \\ 1 & 0 & -1 & 0 \\ 0 & 1 & 1 & -1 \\ 2 & 0 & -1 & -3 \end{vmatrix};$$

$$(c) \begin{vmatrix} 2 & -2 & 1 & 3 \\ 0 & 2 & -1 & -1 \\ 2 & -3 & 2 & 4 \\ 0 & -1 & 1 & 1 \end{vmatrix}; \qquad (d) \det \begin{bmatrix} 2 & 1 & 5 & 2 \\ 2 & 2 & 3 & 0 \\ 2 & 0 & 2 & 1 \\ 3 & 2 & 1 & 0 \end{bmatrix};$$

$$(e) \det \begin{bmatrix} 3 & -3 & 0 & 0 \\ 3 & 2 & 1 & 2 \\ 0 & 2 & 0 & 3 \\ -3 & 0 & -2 & 0 \end{bmatrix}.$$

5. The value of the minor of the entry in the 13th row and 11th column of a 22nd-order matrix is found to be 4. What is the cofactor of this entry? Why?

6. The cofactor of a_{ij} was defined as the product of $(-1)^{i+j}$ and the minor of a_{ij}. The quantity $(-1)^{i+j}$ may be found by the so-called "checker board rule"

$$\begin{vmatrix} + & - & + & - & + & \cdots \\ - & + & - & + & - & \cdots \\ + & - & + & - & + & \cdots \\ - & + & - & + & - & \cdots \\ \cdot & \cdot & \cdot & \cdot & \cdot & \cdot \\ \cdot & \cdot & \cdot & \cdot & \cdot & \cdot \\ \cdot & \cdot & \cdot & \cdot & \cdot & \cdot \end{vmatrix}.$$

The sign in the upper left-hand corner must be positive $(i + j = 1 + 1 = 2)$. If movement is made either horizontally or vertically (but not diagonally) the signs must alternate. By this rule determine the signs which should precede the minors in evaluating the cofactors of a_{41}, a_{24}, and a_{34} in

$$\begin{vmatrix} a_{11} & a_{12} & a_{13} & a_{14} \\ a_{21} & a_{22} & a_{23} & a_{24} \\ a_{31} & a_{32} & a_{33} & a_{34} \\ a_{41} & a_{42} & a_{43} & a_{44} \end{vmatrix}.$$

7. (*a*) Evaluate

$$\begin{vmatrix} a_{11} & a_{12} & a_{13} \\ 0 & a_{22} & a_{23} \\ 0 & 0 & a_{33} \end{vmatrix}.$$

(*b*) Evaluate

$$\begin{vmatrix} a_{11} & 0 & 0 & 0 & \cdots & 0 \\ a_{21} & a_{22} & 0 & 0 & \cdots & 0 \\ a_{31} & a_{32} & a_{33} & 0 & \cdots & 0 \\ \cdot & \cdot & \cdot & \cdot & & \cdot \\ \cdot & \cdot & \cdot & \cdot & & \cdot \\ \cdot & \cdot & \cdot & \cdot & & \cdot \\ a_{n1} & a_{n2} & a_{n3} & a_{n4} & \cdots & a_{nn} \end{vmatrix}$$

(which has only zeros above main diagonal).

8. Without expanding evaluate

$$\begin{vmatrix} 2 & 4 & 6 & 4 \\ 0 & 4 & 6 & 9 \\ 2 & 1 & 4 & 0 \\ 1 & 2 & 3 & 2 \end{vmatrix}.$$

by making use of two theorems of this chapter.

9. Given that $\det A = 8$, and that B is the same matrix as A except that the first and fourth rows have been interchanged. What is the value of $\det B$? Justify your answer.

10. Does $2 \det A = \det (2A)$? Why? (Assume that the order of A is greater than 1.)

11. If $\det [a_{ij}]_{(3,3)} = 4$, find $\det 3[a_{ij}]_{(3,3)}$. Give reasons for your answer.

12. Without expanding show that

$$\begin{vmatrix} x^2 - y^2 & x + y & x \\ x - y & 1 & 1 \\ x - y & 1 & y \end{vmatrix} = 0.$$

Give reasons for your steps.

13. (a) Prove Theorem 3.13.   (b) Prove Theorem 3.15.

14. Prove the following theorem: If any two parallel lines of a matrix A are proportional, then $\det A = 0$.

15. Prove that the converse of the theorem in the preceding exercise is not true.

16. Prove that

$$\begin{vmatrix} a & b & c \\ d & e & f \\ g + h & i + j & k + l \end{vmatrix} = \begin{vmatrix} a & b & c \\ d & e & f \\ g & i & k \end{vmatrix} + \begin{vmatrix} a & b & c \\ d & e & f \\ h & j & l \end{vmatrix}.$$

17. Prove that a determinant of order 3 can be evaluated by the diagonal line method which is demonstrated in Figure 3.4. Then use this method to evaluate $\det \begin{bmatrix} 1 & 3 & 3 \\ 2 & 2 & 1 \\ 4 & 1 & 1 \end{bmatrix}$. The reader should be cautioned that this method is not valid if the order of the determinant is greater than three.

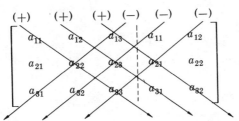

$$|A| = -a_{13}a_{22}a_{31} - a_{11}a_{23}a_{32} - a_{12}a_{21}a_{33} + a_{11}a_{22}a_{33} + a_{12}a_{23}a_{31} + a_{13}a_{21}a_{32}$$

Fig. 3.4

18. In each of the following systems, find each unknown by Cramer's rule.

(a) $\begin{cases} 2x - 2y = 5, \\ x - 4y = 7. \end{cases}$

(b) $\begin{cases} x + y + z = 2, \\ 2x + 3y + 4z = 3, \\ x - 2y - z = 1. \end{cases}$

19. In Example 11 of this section, verify that the formula given for the area is equal to the expansion of the given determinant.

NEW VOCABULARY

§3.1 unary operation	§3.4 determinant of a matrix
§3.2 transpose of a matrix	§3.4 order of a determinant
§3.2 symmetric matrix	§3.4 minor
§3.2 skew-symmetric matrix	§3.4 cofactor
§3.2 conjugate of a matrix	§3.4 expansion about a row
§3.2 Hermitian matrix	or column
§3.2 skew-Hermitian matrix	§3.4 Cramer's rule
§3.3 trace of a matrix	§3.4 triangular matrix
	§3.4 diagonal matrix

Part II:

Systems of Linear Equations

Systems and Rank

4.1 THE NATURE OF SOLUTIONS

As we discovered in Section 2.4, a system of m linear equations in n unknowns

$$\begin{cases} a_{11}x_1 + a_{12}x_2 + \cdots + a_{1n}x_n = b_1, \\ a_{21}x_1 + a_{22}x_2 + \cdots + a_{2n}x_n = b_2, \\ \phantom{a_{21}x_1 + a_{22}x_2 + \cdots + a_{2n}x_n} \cdot \\ \phantom{a_{21}x_1 + a_{22}x_2 + \cdots + a_{2n}x_n} \cdot \\ \phantom{a_{21}x_1 + a_{22}x_2 + \cdots + a_{2n}x_n} \cdot \\ a_{m1}x_1 + a_{m2}x_2 + \cdots + a_{mn}x_n = b_m, \end{cases}$$

can be expressed as a single matrix equation

$$AX = B,$$

where

$$A = \begin{bmatrix} a_{11} & a_{12} & \cdots & a_{1n} \\ a_{21} & a_{22} & \cdots & a_{2n} \\ \cdot & \cdot & & \cdot \\ \cdot & \cdot & & \cdot \\ \cdot & \cdot & & \cdot \\ a_{m1} & a_{m2} & \cdots & a_{mn} \end{bmatrix}, \quad X = \begin{bmatrix} x_1 \\ x_2 \\ \cdot \\ \cdot \\ \cdot \\ x_n \end{bmatrix}, \quad \text{and} \quad B = \begin{bmatrix} b_1 \\ b_2 \\ \cdot \\ \cdot \\ \cdot \\ b_m \end{bmatrix}.$$

Any vector X which satisfies this matrix equation is called a *solution* of the system. *In Part II (Chapters 4, 5, and 6) A, X, and B are real matrices unless otherwise stated.* When $B = 0$, the system is said to be *homogeneous*, and when $B \neq 0$ we say that the system is *non-homogeneous*.

Such a linear system may have exactly one solution, an infinite number of solutions, or no solution at all. Geometrically, these three cases are illustrated, respectively, with three equations and three unknowns in Figure 4.1 (a) through (c); each linear equation is represented by a plane.

71

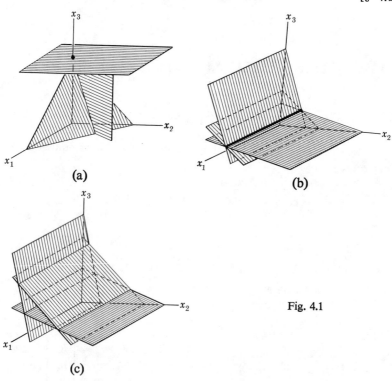

(a)

(b)

(c)

Fig. 4.1

In order to forecast the nature of the solution of a given system of linear equations, we define the **rank** of a matrix and then state two useful theorems.

Definition 4.1. *The **rank** of a matrix is the order of the largest square submatrix whose determinant is not zero. If the entries of the matrix are all zeros, then the rank is zero.*

Example 1. Given

$$A = \begin{bmatrix} 3 & 2 & 1 \\ 0 & 4 & 5 \\ 3 & 6 & 6 \end{bmatrix}.$$

The rank of A is not 3 because det $A = 0$; the rank of A is 2 because we can find a 2 by 2 submatrix whose determinant is not zero.

Example 2. Given

$$B = \begin{bmatrix} -2 & 1 & 3 & 4 \\ 0 & 1 & 1 & 2 \\ 1 & 3 & 4 & 7 \end{bmatrix}.$$

Since there are no submatrices of order 4, the rank is at most 3. To find whether or not it is 3, we must start calculating the determinants of the four 3 by 3 submatrices; as soon as we find a submatrix (if we do) whose determinant is *not* zero, we can stop our calculations with the assurance that the rank of B is 3. Notice that the submatrix resulting from a deletion of the 2nd column has a nonzero determinant, hence the rank of B is 3.

Systems that have one or more solutions are called **consistent** systems, and systems that do not have a solution are called **inconsistent** systems. The following theorem allows us to distinguish between these two types of systems.

Theorem 4.1. *A system of linear equations $AX = B$ is consistent if and only if the rank of the augmented matrix is equal to the rank of the coefficient matrix. This common value of the rank of these two matrices, if it exists, is denoted by r and is called the rank of the system.*

Proof. See Theorem A.6, page 157.

Theorem 4.2. *A consistent system of linear equations $AX = B$ in n unknowns has a unique solution if and only if $r = n$.*

Proof. See Theorem A.7, page 158.

The statements of these two theorems are illustrated in the chart of Figure 4.2.

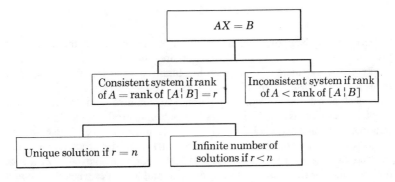

Fig. 4.2

Example 3. The system $AX = B$, or

$$\begin{cases} x_1 + 3x_2 + x_3 = 2, \\ x_1 - 2x_2 - 2x_3 = 3, \\ 2x_1 + x_2 - x_3 = 6, \end{cases}$$

in which $n = m = 3$, is *inconsistent* because the rank of $[A \mid B]$ is 3, and the rank of A is 2.

Example 4. The system $AX = B$, or

$$\begin{cases} x_1 + x_2 - 4x_3 = 2, \\ x_1 - 2x_2 + x_3 = 1, \\ x_1 + x_2 + x_3 = 0, \\ 2x_1 - x_2 + 2x_3 = 1, \end{cases}$$

in which $m = 4$, $n = 3$, is *consistent* because the rank of $[A \mid B]$ is 3, and the rank of A is 3. Therefore, $r = 3$. The system has a unique solution because the rank of the system equals the number of unknowns, that is, $r = n$.

Example 5. The system $AX = B$, or

$$\begin{cases} 3x_1 - x_2 - 2x_3 = 8, \\ 2x_1 - 2x_2 - x_3 = 2, \\ x_1 + x_2 - x_3 = 6, \end{cases}$$

where $m = n = 3$, is *consistent* because the rank of $[A \mid B]$ is 2, and the rank of A is 2. The solution is not unique (that is, there are an infinite number of solutions) because $r = 2$ and $n = 3$.

APPLICATIONS

Example 6. Suppose that a certain company produces three different products: R, S, and T. This output requires the services of two groups of workers; the members of one group are highly trained technicians, and the other group consists of unskilled laborers. Product R requires a day's work from each of 5 technicians and 5 laborers for each unit produced; product S needs 10 technicians and 10 laborers for each unit produced, and product T requires 2 technicians and 4 laborers for each unit produced. The company wants to know how many units of each product to produce each day in order to keep each of its 100 technicians and 150 laborers employed. Let x_1, x_2, and x_3 stand for the number of units of R, S, and T, respectively, that should be manufactured each day to satisfy the stated conditions. Mathematically the problem becomes

$$\begin{cases} 5x_1 + 10x_2 + 2x_3 = 100, \\ 5x_1 + 10x_2 + 4x_3 = 150. \end{cases}$$

There is no unique answer. Subtracting the members of the first equation from those of the second we find $x_3 = 25$. If this value is substituted into either equation, then reduction of the equation yields $x_1 = 10 - 2x_2$. The possible integral values of x_2 are then 0, 1, 2, 3, 4, and 5. Thus, with $x_3 = 25$ we have the following values for x_1 and x_2.

x_1	10	8	6	4	2	0
x_2	0	1	2	3	4	5

To choose among these possibilities the company must use other considerations. Notice that in this problem the components of the solution vector X have to be nonnegative integers rather than real numbers; this explains the reason for obtaining six solutions rather than an infinite number of solutions.

Example 7. A system of linear equations in which there are more equations than unknowns can be obtained from the electrical circuit shown in Figure 4.3. Let i_1, i_2, and i_3 represent the currents, measured in amperes, in their respective parts of the circuit. Using Kirchhoff's rule (named for Gustav Robert Kirchhoff, German, 1824–1887) for relating the voltages, resistances, and currents of a given circuit we obtain the following system of linear equations

$$\begin{cases} i_1 - i_2 - i_3 = 0, \\ 5i_1 + 20i_3 = 50, \\ 10i_2 - 20i_3 = 30, \\ 5i_1 + 10i_2 = 80, \end{cases}$$

The rank of the coefficient matrix is 3, the rank of the augmented matrix is 3, and the number of unknowns is 3, hence there is a unique solution; that is, there is only one vector, with components $i_1 = 6$, $i_2 = 5$, $i_3 = 1$, which satisfies all the equations simultaneously.

Example 8. Suppose that a certain manufacturer produces two brands of fertilizer — P and Q. Each pound of brand P contains 20 units of chemical A and 10 units of chemical B; each pound of brand Q contains 30 units of chemical A and 50 units of chemical B. Suppose there is a limited supply of 1,200 units of chemical A and 900 units of chemical B, but that all other

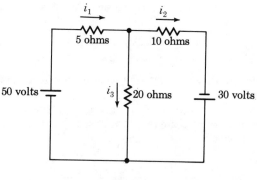

Fig. 4.3

ingredients are in ample supply. How many pounds of each brand of fertilizer can be produced within the restrictions imposed by the limited supply of chemicals?

Solution: If we let x_1 represent the number of pounds of P produced and x_2 represent the number of pounds of Q produced, then the problem posed above can be restated as a system of linear inequalities

$$\begin{cases} 20x_1 + 30x_2 \leq 1200, \\ 10x_1 + 50x_2 \leq 900, \\ \quad x_1 \qquad\quad \geq \quad 0, \\ \qquad\quad x_2 \geq \quad 0. \end{cases}$$

In this particular problem, the solution set is a region bounded by the system of lines

$$\begin{cases} 20x_1 + 30x_2 = 1200, \\ 10x_1 + 50x_2 = 900, \\ \quad x_1 \qquad\quad = \quad 0, \\ \qquad\quad x_2 = \quad 0. \end{cases}$$

This system of equations is inconsistent (notice that the rank of the augmented matrix is 3 and the rank of the coefficient matrix is 2); the graph of the system of equations is shown in Figure 4.4. The solution set, called the *feasible set,* for the original system of inequalities is shaded in Figure 4.4; this is the region common to the four half-planes that satisfy the four given inequalities. Feasible amounts of the two brands of fertilizers that can be produced are represented by the coordinates of any point in the feasible set, for example, $x_1 = 40$ (pounds of P) and $x_2 = 10$ (pounds of Q), or $x_1 = 20$ (pounds of P) and $x_2 = 14$ (pounds of Q).

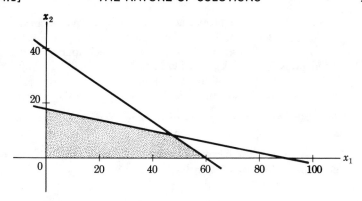

Fig. 4.4

This shaded region, or feasible set, is an example of a *polyhedral convex set*. Before formally defining a polyhedral convex set we point out that the solution set of a single inequality of the form

$$c_1x_1 + c_2x_2 + \cdots + c_nx_n \leq d$$

is called a *closed half-space*. Also, the *intersection* of n closed half-spaces is the set of points common to all n of the half-spaces; it is the set of points whose coordinates satisfy simultaneously the n corresponding inequalities.

Definition 4.2. *The intersection of a finite number of closed half-spaces is called a polyhedral convex set.*

The application demonstrated in Example 8 is developed further in Exercise 9 of Section 6.2.

EXERCISES

1. Find the ranks of:

$$(a) \begin{bmatrix} 1 & 2 & -1 \\ 4 & 1 & 5 \\ 3 & -1 & 6 \end{bmatrix}; \quad (b) \begin{bmatrix} 3 & 1 & 4 & 4 \\ 2 & 1 & 3 & 1 \\ 0 & 2 & 2 & 0 \end{bmatrix}; \quad (c) \begin{bmatrix} 2 & 4 \\ -4 & -8 \\ 1 & 2 \end{bmatrix};$$

$$(d) \begin{bmatrix} 0 & 0 & 0 \\ 0 & 0 & 0 \end{bmatrix}; \quad (e) \begin{bmatrix} 0 & 0 & 2 \\ 0 & 0 & 0 \end{bmatrix}.$$

2. Are the following statements true or false? Give your reasons.

(a) If $[a_{ij}]_{(4,5)}$ has rank 3, then the determinant of every fourth-order submatrix is 0; and, conversely, if every such fourth-order determinant is 0, the rank of the matrix is 3.

(b) The rank of A is equal to the rank of A^T.

(c) If 3 by 3 A is skew-symmetric, its rank is less than 3.

3. What is the largest possible rank of an m by n matrix under the assumption that $m > n$?

4. Under what condition is the rank of the following matrix 3? Is it possible for the rank to be 1? Why?

$$\begin{bmatrix} 2 & 4 & 2 \\ 2 & 1 & 2 \\ 1 & 0 & x \end{bmatrix}.$$

5. Let $A = [a_{ij}]_{(4,4)}$.

(a) If $\det A = 0$, what can be said about the rank of A?

(b) If $\det A \neq 0$, what can be said about the rank of A?

6. Determine whether the following systems are inconsistent, consistent with a unique solution, or consistent with an infinite number of solutions. Find r, if it exists, and state the relation between r and n.

(a) $\begin{cases} x_1 + 3x_2 + x_3 = 4, \\ x_1 + x_2 - x_3 = 1, \\ 2x_1 + 4x_2 = 0. \end{cases}$
(b) $\begin{cases} x_1 - x_2 + 6 = 0, \\ x_1 + 2x_2 - 5 = 0, \\ 3x_1 + 3x_2 - 4 = 0. \end{cases}$

(c) $\begin{cases} x_1 + x_2 + x_3 - 5 = 0, \\ x_1 - x_2 - x_3 - 4 = 0. \end{cases}$
(d) $\begin{bmatrix} 1 & 1 & 1 \\ 1 & -1 & 0 \\ 2 & 0 & 1 \end{bmatrix} \begin{bmatrix} x_1 \\ x_2 \\ x_3 \end{bmatrix} = \begin{bmatrix} 0 \\ 0 \\ 0 \end{bmatrix}.$

(e) $AX = 0$, where $A = \begin{bmatrix} 2 & 1 & 0 \\ 1 & -1 & 1 \\ 0 & 1 & -1 \end{bmatrix}$ and $X = \begin{bmatrix} x_1 \\ x_2 \\ x_3 \end{bmatrix}.$

(f) $AX = B$, where $B = \begin{bmatrix} 2 \\ 0 \\ 4 \end{bmatrix}$, $A = \begin{bmatrix} 1 & 1 & -1 \\ 2 & 2 & -1 \\ 1 & 1 & 0 \end{bmatrix}$, $X = \begin{bmatrix} x_1 \\ x_2 \\ x_3 \end{bmatrix}.$

7. Which of the systems in Exercise 6 are homogeneous?

8. A certain system of linear equations with six unknowns is known to be consistent and to have a unique solution. What can be said about the ranks of the augmented and coefficient matrices? What can be said about the number of equations? Give reasons.

9. The rank of the coefficient matrix of a certain inconsistent system of linear equations with five unknowns is found to be 4. What can be said about the rank of the augmented matrix? What can be said about the number of equations?

10. Suppose we are given 5 linear homogeneous equations in 5 unknowns. If det $A \neq 0$, what can we say about a solution?

11. The system $AX = B$ consists of m linear equations (which may or may not be homogeneous) in n unknowns. Assume that r exists for this system. Why is it always true that $r \leq m$ and $r \leq n$?

NEW VOCABULARY

§4.1   solution of the system
§4.1   homogeneous system
§4.1   nonhomogeneous
           system
§4.1   rank of a matrix
§4.1   consistent system
§4.1   inconsistent system

§4.1   rank of the system
§4.1   feasible set
§4.1   polyhedral convex set
§4.1   closed half-spaces
§4.1   intersection of closed
           half-spaces

Systems with a
Unique Solution

5.1 THE INVERSE MATRIX

The identity matrix I for matrix multiplication has been previously defined as a square matrix with every entry of the main diagonal equal to 1 and all other entries equal to 0; the reason I is called the identity matrix for matrix multiplication is that for any square matrix A

$$AI = IA = A$$

(A and I must be of the same order). If we wish to show that the order of the identity matrix is n, we write I_n.

Definition 5.1. *An inverse matrix A^{-1} of a given square matrix A, if it exists, is a square matrix such that*

$$AA^{-1} = A^{-1}A = I,$$

where I is the identity matrix whose order is the same as that of A.

Two immediate questions come to mind. When does A^{-1} exist? If the inverse does exist, how can we find it? Before these two questions are answered by Theorem 5.1 we must define a few terms.

Definition 5.2. *A square matrix A is **singular** if $\det A = 0$ and **nonsingular** if $\det A \neq 0$.*

Definition 5.3. *Let $A = [a_{ij}]_n$ be a square matrix, where $n \geq 2$. The **cofactor matrix** of A, designated by cof A, is the matrix of order n whose entry in row i and column j is A_{ij}, the cofactor of a_{ij} in A.*

Example 1. Let

$$A = \begin{bmatrix} 2 & 4 & 0 \\ 0 & 2 & 1 \\ 3 & 0 & 2 \end{bmatrix}.$$

$$\text{cof } A = \begin{bmatrix} +\begin{vmatrix} 2 & 1 \\ 0 & 2 \end{vmatrix} & -\begin{vmatrix} 0 & 1 \\ 3 & 2 \end{vmatrix} & +\begin{vmatrix} 0 & 2 \\ 3 & 0 \end{vmatrix} \\ -\begin{vmatrix} 4 & 0 \\ 0 & 2 \end{vmatrix} & +\begin{vmatrix} 2 & 0 \\ 3 & 2 \end{vmatrix} & -\begin{vmatrix} 2 & 4 \\ 3 & 0 \end{vmatrix} \\ +\begin{vmatrix} 4 & 0 \\ 2 & 1 \end{vmatrix} & -\begin{vmatrix} 2 & 0 \\ 0 & 1 \end{vmatrix} & +\begin{vmatrix} 2 & 4 \\ 0 & 2 \end{vmatrix} \end{bmatrix} = \begin{bmatrix} 4 & 3 & -6 \\ -8 & 4 & 12 \\ 4 & -2 & 4 \end{bmatrix}.$$

Also, since det $A = 20 \neq 0$, A is nonsingular.

Definition 5.4. *The **adjoint matrix**, designated by* adj A, *of a square matrix A (of order $n \geq 2$) is the transpose of* cof A.

Example 2. Using the matrix A of the previous example, we have

$$\text{adj } A = (\text{cof } A)^{\text{T}} = \begin{bmatrix} 4 & 3 & -6 \\ -8 & 4 & 12 \\ 4 & -2 & 4 \end{bmatrix}^{\text{T}} = \begin{bmatrix} 4 & -8 & 4 \\ 3 & 4 & -2 \\ -6 & 12 & 4 \end{bmatrix}.$$

We are now in a position to answer the questions we raised earlier about the existence and formation of an inverse of a given matrix with the following key theorem.

Theorem 5.1. *For a square matrix A an inverse A^{-1} exists if and only if A is nonsingular. Moreover, if A^{-1} exists, then*

$$A^{-1} = \frac{1}{|A|} \text{ adj } A.$$

Proof. Since the first assertion contains the double conditional "if and only if" we must prove the two assertions: (1) The nonsingularity of A implies the existence of A^{-1}, and, conversely, (2) the existence of A^{-1} implies that A is nonsingular.

Assertion (1). Assume that A is nonsingular (that is, $|A| \neq 0$), and prove that A^{-1} exists.

STATEMENT

REASON

(1) $A\,(\text{adj}\,A) =$

$$\begin{bmatrix} a_{11} & a_{12} & \cdots & a_{1n} \\ a_{21} & a_{22} & \cdots & a_{2n} \\ \cdot & \cdot & \cdot & \cdot \\ \cdot & \cdot & & \cdot \\ \cdot & \cdot & & \cdot \\ a_{n1} & a_{n2} & \cdots & a_{nn} \end{bmatrix} \begin{bmatrix} A_{11} & A_{21} & \cdots & A_{n1} \\ A_{12} & A_{22} & \cdots & A_{n2} \\ \cdot & \cdot & & \cdot \\ \cdot & \cdot & \cdot & \cdot \\ \cdot & \cdot & \cdot & \cdot \\ A_{1n} & A_{2n} & \cdots & A_{nn} \end{bmatrix}$$

(1) Definition 5.4 (adjoint matrix); Definition 3.10 (determinant); and Theorem 3.12 after matrix multiplication.

$$= \begin{bmatrix} |A| & 0 & 0 & 0 & \cdots & 0 \\ 0 & |A| & 0 & 0 & \cdots & 0 \\ 0 & 0 & |A| & 0 & \cdots & 0 \\ \cdot & & \cdot & \cdot & \cdot & \cdot \\ \cdot & & & \cdot & & \cdot \\ \cdot & & & & \cdot & \cdot \\ 0 & 0 & 0 & 0 & \cdots & |A| \end{bmatrix}.$$

(2) $A\,(\text{adj}\,A) = |A|I_n.$

(2) Definition 2.3 (multiplication of a matrix by a scalar).

(3) $|A| \neq 0.$

(3) Assumption.

(4) $A\,\dfrac{\text{adj}\,A}{|A|} = I_n.$

(4) Statements (1), (2), and (3).

(5) If we define $A_1 = \dfrac{\text{adj}\,A}{|A|}$, then A_1 exists and $AA_1 = I_n.$

(5) Statement (4).

(6) Also $A_1 A = I_n.$

(6) An argument similar to Statements (1) to (5).

(7) $A^{-1} = A_1 = \dfrac{\text{adj}\,A}{|A|}$

(7) Statements (5) and (6) and Definition 5.1.

Thus the first part of the proof is complete. Moreover, we have developed a formula for calculating A^{-1}.

Assertion (2). Assume that A^{-1} exists, and prove that A is nonsingular.

STATEMENT	REASON
(1) $AA^{-1} = A^{-1}A = I_n$.	(1) Assumption that A^{-1} exists satisfying Definition 5.1.
(2) $|A||A^{-1}| = |I_n|$.	(2) Theorem 3.13 $|AA^{-1}| = |A||A^{-1}|$.
(3) $|A| \neq 0$.	(3) If $|A| = 0$, then by Statement (2), $|I_n| = 0$. But $|I_n| = 1$.
(4) A is nonsingular.	(4) Definition 5.2.

The following example illustrates the use of the formula given in Theorem 5.1 for finding the inverse of a given matrix.

Example 3. In Examples 1 and 2 we found that if

$$A = \begin{bmatrix} 2 & 4 & 0 \\ 0 & 2 & 1 \\ 3 & 0 & 2 \end{bmatrix}, \quad \text{then} \quad \text{adj } A = \begin{bmatrix} 4 & -8 & 4 \\ 3 & 4 & -2 \\ -6 & 12 & 4 \end{bmatrix}.$$

We find also that $|A| = 20$; therefore,

$$A^{-1} = \frac{\text{adj } A}{|A|} = \begin{bmatrix} \frac{4}{20} & -\frac{8}{20} & \frac{4}{20} \\ \frac{3}{20} & \frac{4}{20} & -\frac{2}{20} \\ -\frac{6}{20} & \frac{12}{20} & \frac{4}{20} \end{bmatrix}.$$

To check our result, we find that

$$AA^{-1} = A^{-1}A = I_3.$$

Now that we have learned how to determine the existence of an inverse of a square matrix and a method of its calculation when it exists, we ask the question: Is the inverse unique? That is, is there only one inverse for a matrix A?

Theorem 5.2. *If A has an inverse, then A^{-1} is unique.*

Proof. We assume that A has two inverses B and C, and then we show that $B = C$. We therefore assume that $AB = BA = I$ and $AC = CA = I$. It follows then that

$$B = BI = B(AC) = (BA)C = IC = C.$$

(The reasons for the above statements are left as an exercise for the reader to fill in.)

Negative integral exponents for nonsingular matrices can now be defined as follows:

$$A^{-2} = (A^{-1})^2 = A^{-1}A^{-1},$$
$$A^{-3} = (A^{-1})^3 = A^{-1}A^{-1}A^{-1},$$

$$\vdots \qquad \vdots$$

$$A^{-n} = (A^{-1})^n = A^{-1}A^{-1} \ldots A^{-1}A^{-1} \qquad (A^{-1} \text{ occurring } n \text{ times}).$$

APPLICATIONS

★ A primary application of the inverse matrix is in seeking the solutions of systems of linear equations where the coefficient matrix is nonsingular. Consider the system $AX = B$; premultiply both sides of the equation by A^{-1}:

$$A^{-1}(AX) = A^{-1}B,$$
$$(A^{-1}A)X = A^{-1}B,$$
$$IX = A^{-1}B,$$
$$X = A^{-1}B.$$

The inverse matrix is especially useful in those systems where B varies but the coefficient matrix A remains unchanged.

Example 4. The electrical system discussed in Example 7 of Section 4.1, page 75, and shown in Figure 4.3 gives rise to the following linear system:

$$\begin{cases} x_1 - x_2 - x_3 = 0, \\ 5x_1 \qquad\ + 20x_3 = 50, \\ \quad\ 10x_2 - 20x_3 = 30, \end{cases}$$

or

$$AX = B,$$

where X is the electric current vector $(x_1 = i_1, x_2 = i_2, x_3 = i_3)$ and the last two components of B are the electromotive forces. Here,

$$A = \begin{bmatrix} 1 & -1 & -1 \\ 5 & 0 & 20 \\ 0 & 10 & -20 \end{bmatrix},$$

so that

$$|A| = -350 \quad \text{and} \quad \text{cof } A = \begin{bmatrix} -200 & 100 & 50 \\ -30 & -20 & -10 \\ -20 & -25 & 5 \end{bmatrix}.$$

Therefore,

$$A^{-1} = \frac{1}{|A|}(\text{cof } A)^{\mathrm{T}} = \begin{bmatrix} \frac{4}{7} & \frac{3}{35} & \frac{4}{70} \\ -\frac{2}{7} & \frac{2}{35} & \frac{5}{70} \\ -\frac{1}{7} & \frac{1}{35} & -\frac{1}{70} \end{bmatrix}.$$

Since

$$B = \begin{bmatrix} 0 \\ 50 \\ 30 \end{bmatrix},$$

$$X = A^{-1}B = \begin{bmatrix} \frac{4}{7} & \frac{3}{35} & \frac{4}{70} \\ -\frac{2}{7} & \frac{2}{35} & \frac{5}{70} \\ -\frac{1}{7} & \frac{1}{35} & -\frac{1}{70} \end{bmatrix} \begin{bmatrix} 0 \\ 50 \\ 30 \end{bmatrix} = \begin{bmatrix} \frac{150}{35} + \frac{60}{35} \\ \frac{100}{35} + \frac{75}{35} \\ \frac{50}{35} - \frac{15}{35} \end{bmatrix} = \begin{bmatrix} 6 \\ 5 \\ 1 \end{bmatrix}.$$

If both electromotive forces of the circuit were changed to 70 volts the solution would be

$$X = A^{-1} \begin{bmatrix} 0 \\ 70 \\ 70 \end{bmatrix} = \begin{bmatrix} 10 \\ 9 \\ 1 \end{bmatrix}.$$

Example 5. The management of a certain company is confronted with a decision. One of its factories uses two different machines M and N to manufacture two different products P and Q. Machine M can operate 12 hours per day, and machine N can operate 16 hours per day; the rest of the day is used for the maintenance of the machines. To produce one unit of product P requires that work be done for 2 hours a day by machine M and for 1 hour a day by machine N. Each unit of product Q requires that machine M work 2 hours a day and that machine N work 3 hours a day. Find the number of units of each product that the factory should make in a day in order to keep the machines working to capacity. Also, what is the effect on production if we buy more of each type of machine?

Solution. Let x_1 be the number of units of product P that are produced; let x_2 be the number of units of product Q that are produced. Machine M then spends $2x_1$ hours on product P and $2x_2$ hours on product Q. If machine M operates full time we have the equation

$$2x_1 + 2x_2 = 12.$$

Similarly if machine N operates full time we get the equation

$$x_1 + 3x_2 = 16.$$

The two simultaneous equations can be expressed as the matrix equation

$$AX = B,$$

where

$$A = \begin{bmatrix} 2 & 2 \\ 1 & 3 \end{bmatrix} \quad \text{and} \quad B = \begin{bmatrix} 12 \\ 16 \end{bmatrix}.$$

Therefore,

$$X = A^{-1}B = A^{-1}\begin{bmatrix} 12 \\ 16 \end{bmatrix}.$$

By the adjoint method we easily find that

$$A^{-1} = \begin{bmatrix} \frac{3}{4} & -\frac{1}{2} \\ -\frac{1}{4} & \frac{1}{2} \end{bmatrix};$$

hence,

$$X = \begin{bmatrix} \frac{3}{4} & -\frac{1}{2} \\ -\frac{1}{4} & \frac{1}{2} \end{bmatrix}\begin{bmatrix} 12 \\ 16 \end{bmatrix} = \begin{bmatrix} 1 \\ 5 \end{bmatrix}.$$

Therefore one unit of product P and five units of product Q should be produced to satisfy the initial stated conditions of the problem.

Now if the management buys more machines, the resulting change in production can be determined by simply changing B and recalculating $A^{-1}B$. Notice that because A^{-1} does not change, it is relatively easy to find the new solution since it is in the calculation of the inverse matrix that most of the work was done. Suppose the management buys an extra machine M. This increases the total capacity of machines M to 24 hours; thus,

$$X = A^{-1}B = A^{-1}\begin{bmatrix} 24 \\ 16 \end{bmatrix} = \begin{bmatrix} \frac{3}{4} & -\frac{1}{2} \\ -\frac{1}{4} & \frac{1}{2} \end{bmatrix}\begin{bmatrix} 24 \\ 16 \end{bmatrix} = \begin{bmatrix} 10 \\ 2 \end{bmatrix}.$$

Example 6. Consider a certain corporation that has three fields of operation; it mines coal, produces gasoline, and generates electricity. Each of these *activities* makes use of varying amounts of the three products. Suppose that in order to produce *one unit* of coal the corporation consumes

> 0 units of coal,
> 1 unit of gasoline,
> 1 unit of electricity;

to produce *one unit* of gasoline the corporation consumes

0 units of coal,

$\frac{1}{5}$ unit of gasoline,

$\frac{2}{5}$ unit of electricity;

and to produce *one unit* of electricity the corporation consumes

$\frac{1}{5}$ unit of coal,

$\frac{2}{5}$ unit of gasoline,

$\frac{1}{5}$ unit of electricity.

These three vectors form what is known as a *consumption matrix*

$$C = \begin{array}{c} \\ \\ \\ \\ \end{array} \begin{array}{ccc} \text{COAL} & \text{GAS} & \text{ELECTRICITY} \\ \begin{bmatrix} 0 & 0 & \frac{1}{5} \\ 1 & \frac{1}{5} & \frac{2}{5} \\ 1 & \frac{2}{5} & \frac{1}{5} \end{bmatrix} & & \begin{array}{l} \text{COAL CONSUMED} \\ \text{GAS CONSUMED} \\ \text{ELECTRICITY CONSUMED.} \end{array} \end{array}$$

For a given time interval, let x_1 be the number of units of coal produced, let x_2 be the number of units of gas produced, and let x_3 be the number of units of electricity produced. Thus

$$X = \begin{bmatrix} x_1 \\ x_2 \\ x_3 \end{bmatrix}$$

is the production vector. The product CX shows the internal consumption necessary for this desired level of production. Naturally, the corporation wishes to produce more than its internal needs require. Suppose it has an order for 100 units of each product. How much of each product should the corporation produce in order to meet this demand? The solution is found by considering the nonnegative solution of the matrix equation $X - CX = D$, if such a solution exists, where $D = \begin{bmatrix} 100 \\ 100 \\ 100 \end{bmatrix}$. This equation can be rewritten as

$$(I - C)X = D \qquad \text{or} \qquad X = (I - C)^{-1}D$$

if $(I - C)^{-1}$ exists. For our data this becomes

$$X = \left(\begin{bmatrix} 1 & 0 & 0 \\ 0 & 1 & 0 \\ 0 & 0 & 1 \end{bmatrix} - \begin{bmatrix} 0 & 0 & \frac{1}{5} \\ 1 & \frac{1}{5} & \frac{2}{5} \\ 1 & \frac{2}{5} & \frac{1}{5} \end{bmatrix} \right)^{-1} \begin{bmatrix} 100 \\ 100 \\ 100 \end{bmatrix} = \begin{bmatrix} 300 \\ 1000 \\ 1000 \end{bmatrix}.$$

Once a corporation's internal consumption is established, then $(I - C)^{-1}$ (if it exists) can be calculated once and for all; with each new demand vector, the problem then amounts to evaluating a simple product of matrices.

EXERCISES

1. Calculate the adjoint matrices for the following matrices:

$$A = \begin{bmatrix} 2 & 3 \\ -4 & 1 \end{bmatrix}; \quad B = \begin{bmatrix} 1 & 3 & 0 \\ 2 & 1 & 0 \\ 0 & 1 & -1 \end{bmatrix}; \quad C = \begin{bmatrix} 1 & 0 & 0 & 1 \\ 0 & 0 & 1 & 0 \\ 0 & 2 & 0 & 0 \\ 0 & 0 & 0 & 2 \end{bmatrix}.$$

2. Making use of the answers to the preceding exercise, calculate the inverses of those matrices.

3. Calculate the inverses of the following matrices if possible. Check by $AA^{-1} = A^{-1}A = I$.

$$A = \begin{bmatrix} 2 & -1 \\ 4 & 3 \end{bmatrix}; \quad B = \begin{bmatrix} 4 & 2 \\ 2 & 1 \end{bmatrix}; \quad C = \begin{bmatrix} 2 & 0 & 3 \\ -1 & 0 & 2 \\ 0 & 1 & 1 \end{bmatrix}; \quad D = \begin{bmatrix} 3 & 2 & 1 \\ 2 & -1 & -1 \\ 1 & 4 & 0 \end{bmatrix}.$$

4. What can be said about the rank of a nonsingular matrix of order 3?

5. If cof A is symmetric, how do cof A and adj A compare?

6. State why it is impossible for a matrix that is not square to have an inverse.

7. If $A = \begin{bmatrix} 3 & 2 \\ 0 & 1 \end{bmatrix}$, find A^{-2} and A^{-3}.

8. Prove that if det $A \neq 0$, then det $(A^{-1}) = (\det A)^{-1}$. (*Hint:* Make use of Theorem 3.14.) Note the different meaning of the exponents in this exercise.

9. Prove the theorem: If A is nonsingular, then $(A^{-1})^{-1} = A$.

10. Prove the theorem: If A is nonsingular, then $(A^{T})^{-1} = (A^{-1})^{T}$. (*Hint:* Use Theorem 5.2.)

11. Prove the theorem: If A and B are nonsingular, then $(AB)^{-1} = B^{-1}A^{-1}$. (*Hint:* Use Theorem 5.2.)

12. Prove that if A and B are square and if $AB = I$, then $BA = AB = I$.

13. Let A, B, and C be square matrices of the same order. If A is nonsingular, prove that $BA = CA$ implies that $B = C$.

14. Rework Example 4 of this section with the electromotive force increased to 100 volts in each battery.

15. Rework Example 5 of this section with the company having two of each type of machine.

16. In Example 6 of this section what is the internal consumption?

17. For a square matrix A, does $(\text{cof } A)^T = \text{cof }(A^T)$?

5.2 ELEMENTARY OPERATIONS

The reader is probably familiar with the addition–subtraction method of solving a system of n linear equations in n unknowns. We perform the operations necessary to solve the following system of linear equations and simultaneously we perform the same operations on the augmented matrix of this system. Our guiding purpose in the following manipulations is to reduce the coefficient matrix to triangular form (for a definition of a triangular matrix, see Definition 3.11):

$$\begin{cases} x_2 + x_3 = 1, \\ x_2 - x_3 = 5, \\ x_1 + x_2 + x_3 = 2. \end{cases} \qquad \begin{bmatrix} 0 & 1 & 1 & \vdots & 1 \\ 0 & 1 & -1 & \vdots & 5 \\ 1 & 1 & 1 & \vdots & 2 \end{bmatrix}$$

Interchange Row 1 and Row 3;

$$\begin{cases} x_1 + x_2 + x_3 = 2, \\ x_2 - x_3 = 5, \\ x_2 + x_3 = 1. \end{cases} \qquad \begin{bmatrix} 1 & 1 & 1 & \vdots & 2 \\ 0 & 1 & -1 & \vdots & 5 \\ 0 & 1 & 1 & \vdots & 1 \end{bmatrix}$$

Multiply Row 2 by (-1) and add to Row 3;

$$\begin{cases} x_1 + x_2 + x_3 = 2, \\ x_2 - x_3 = 5, \\ 2x_3 = -4. \end{cases} \qquad \begin{bmatrix} 1 & 1 & 1 & \vdots & 2 \\ 0 & 1 & -1 & \vdots & 5 \\ 0 & 0 & 2 & \vdots & -4 \end{bmatrix}$$

Multiply Row 3 by $\frac{1}{2}$;

$$\begin{cases} x_1 + x_2 + x_3 = 2, \\ x_2 - x_3 = 5, \\ x_3 = -2. \end{cases} \qquad \begin{bmatrix} 1 & 1 & 1 & \vdots & 2 \\ 0 & 1 & -1 & \vdots & 5 \\ 0 & 0 & 1 & \vdots & -2 \end{bmatrix}$$

These four systems of equations are said to be **equivalent** because they have the same solution. From the last system it is an easy matter to use substitution to determine the solution $x_1 = 1$, $x_2 = 3$, $x_3 = -2$. Notice that with proper interpretation, the operations performed on the augmented matrix accomplish the same result as manipulating the equations themselves. This treatment of the augmented matrix is analogous to synthetic division in that we deal only with coefficients and the position of the entry determines the corresponding variable. The procedure demonstrated above is often referred to as the *Gauss elimination method,* (named for Karl Friedrich Gauss, German, 1777–1855), the *echelon method,* or *triangularization.*

The three operations used in the illustration above are called *elementary row operations* and now are stated formally.

Definition 5.5. *The following three operations:*
 (1) *Interchange of any two rows (denoted by $R_i \leftrightarrow R_j$);*
 (2) *Multiplication of any row by a nonzero scalar (denoted by kR_i);*
 (3) *Addition to any row of a scalar multiple of another row (denoted by $kR_i + R_j$);*
are called elementary row operations on a matrix.

The analogous *elementary column operations* are similarly defined by replacing the word "row" by "column" and the letter R by C throughout the preceding definition. *An elementary operation is any operation that is either an elementary row operation or an elementary column operation.* If a matrix A can be transformed into a matrix B by means of one or more elementary operations, we write $A \sim B$ and say that A is equivalent to B. In particular, we say that A is *row equivalent* (or *column equivalent*) to B if only elementary row (or column) operations are involved in the transformation.

Elementary operations can be used to reduce a nonzero matrix to what is called *normal form.* (You are asked to prove this assertion in Exercise 14.)

Definition 5.6. *When a matrix has been reduced by elementary operations to one of the following forms*

$$\begin{bmatrix} I_r \\ -- \\ 0 \end{bmatrix}, \quad [I_r \mid 0], \quad \begin{bmatrix} I_r & \mid & 0 \\ --- & \mid & --- \\ 0 & \mid & 0 \end{bmatrix}, \quad or \quad I_r,$$

we say that the matrix has been reduced to normal form.

Example 1.   Reduce

$$A = \begin{bmatrix} 2 & 1 & 4 \\ 1 & 0 & 4 \\ -2 & 0 & 2 \end{bmatrix}$$

to normal form.

Solution.

$A \underset{C_1 \longleftrightarrow C_2}{\sim} \begin{bmatrix} 1 & 2 & 4 \\ 0 & 1 & 4 \\ 0 & -2 & 2 \end{bmatrix}$ The purpose of this step is to obtain a 1 in the a_{11} position by operation (1) of elementary operations; we are lucky enough to get zeros elsewhere in the first column.

$\underset{\substack{-2C_1+C_2 \\ -4C_1+C_3}}{\sim} \begin{bmatrix} 1 & 0 & 0 \\ 0 & 1 & 4 \\ 0 & -2 & 2 \end{bmatrix}$ Here, except in the first column, we obtain zeros in the first row by application of column operation (3) twice, without disturbing the rest of the matrix.

$\underset{2R_2+R_3}{\sim} \begin{bmatrix} 1 & 0 & 0 \\ 0 & 1 & 4 \\ 0 & 0 & 10 \end{bmatrix}$ For the lower right 2 by 2 submatrix, using row operation (3), we begin to repeat the process of obtaining zero entries.

$\underset{-4C_2+C_3}{\sim} \begin{bmatrix} 1 & 0 & 0 \\ 0 & 1 & 0 \\ 0 & 0 & 10 \end{bmatrix}$ By column operation (3) we were able to eliminate the a_{23} entry.

$\underset{\frac{1}{10}R_3}{\sim} \begin{bmatrix} 1 & 0 & 0 \\ 0 & 1 & 0 \\ 0 & 0 & 1 \end{bmatrix}$ The matrix A is now in normal form; the last step was completed by row operation (2).

Since $|A| \neq 0$, what we did in Example 1 could have been done with row operations only, but not as easily. There are many paths by which one may travel to attain the normal form. We suggest that the reader establish some suitable, systematic procedure and stick to it. We will use the normal form of a matrix as a possible shortcut for finding the rank of a matrix. We justify the procedure by the following theorem.

Theorem 5.3.   *Equivalent matrices have the same rank.*

Proof.   For an outline of the proof see Theorem A.8, page 159.

Thus in the preceding example the matrix A has rank 3 because it is equivalent to I_3, which obviously has rank 3. In general, the rank of the matrix in Definition 5.6 is r.

Any single elementary operation on a matrix can also be accomplished by multiplying the matrix by a suitably chosen *elementary matrix*.[1]

Definition 5.7. *An **elementary matrix** is a matrix that can be obtained from the identity matrix I by an elementary operation.*

Example 2. The first and second rows of a matrix $A = [a_{ij}]_{(3,2)}$ may be interchanged by premultiplying A by

$$E = \begin{bmatrix} 0 & 1 & 0 \\ 1 & 0 & 0 \\ 0 & 0 & 1 \end{bmatrix}.$$

Note that the desired elementary matrix E is found by simply performing the specified elementary operation on an identity matrix of the appropriate order.

$$EA = \begin{bmatrix} 0 & 1 & 0 \\ 1 & 0 & 0 \\ 0 & 0 & 1 \end{bmatrix} \begin{bmatrix} a_{11} & a_{12} \\ a_{21} & a_{22} \\ a_{31} & a_{32} \end{bmatrix} = \begin{bmatrix} a_{21} & a_{22} \\ a_{11} & a_{12} \\ a_{31} & a_{32} \end{bmatrix}.$$

The first and second columns are interchanged by postmultiplying A by

$$E = \begin{bmatrix} 0 & 1 \\ 1 & 0 \end{bmatrix}.$$

$$AE = \begin{bmatrix} a_{11} & a_{12} \\ a_{21} & a_{22} \\ a_{31} & a_{32} \end{bmatrix} \begin{bmatrix} 0 & 1 \\ 1 & 0 \end{bmatrix} = \begin{bmatrix} a_{12} & a_{11} \\ a_{22} & a_{21} \\ a_{32} & a_{31} \end{bmatrix}.$$

Example 3. To add k times the second row to the first row of A, we premultiply by

$$E = \begin{bmatrix} 1 & k & 0 \\ 0 & 1 & 0 \\ 0 & 0 & 1 \end{bmatrix}.$$

$$\begin{bmatrix} 1 & k & 0 \\ 0 & 1 & 0 \\ 0 & 0 & 1 \end{bmatrix} \begin{bmatrix} a_{11} & a_{12} \\ a_{21} & a_{22} \\ a_{31} & a_{32} \end{bmatrix} = \begin{bmatrix} a_{11} + ka_{21} & a_{12} + ka_{22} \\ a_{21} & a_{22} \\ a_{31} & a_{32} \end{bmatrix}.$$

To add k times the second column to the first column of A we postmultiply by

$$E = \begin{bmatrix} 1 & 0 \\ k & 1 \end{bmatrix}.$$

---

[1] A justification of this statement may be found in the text, C. G. Cullen, *Matrices and Linear Transformations*, Reading, Massachusetts, Addison-Wesley, 1966. Pp. 30–31.

By definition, elementary matrices are obtained by performing a single elementary operation on an identity matrix. We can therefore conclude that any elementary matrix E is nonsingular and thus has an inverse. (Why?) This inverse E^{-1} when applied to a matrix undoes whatever E did. That is, if $EA = B$, then premultiplication of both sides by E^{-1} yields

$$E^{-1}EA = E^{-1}B,$$
$$IA = E^{-1}B,$$
$$A = E^{-1}B.$$

APPLICATIONS

Systems of linear equations are useful in many fields of study. With high-speed computers now available to do the labor, the Gauss elimination method of solution has become very important. It is an alternative method of solving the applied problems of the last section; moreover, it is a method which is applicable when an inverse matrix does not exist (this is demonstrated in the next chapter).

So far we have demonstrated the use of elementary operations in the calculation of rank and in the Gauss elimination method. Elementary operations also are used in the simplex method of linear programming,[2] a new area of mathematics of interest in solving problems in agriculture, engineering, business, and many other fields. The next example points out a fourth use.

★ **Example 4.**   The purpose of this example is to illustrate, and then to justify, a reasonably efficient alternative method for finding the inverse of a nonsingular matrix. First, we augment the nonsingular matrix, whose inverse we wish to find, by I; that is, we write $[A \mid I]$. Then we use elementary row operations to transform this matrix to the form $[I \mid P]$; we shall then prove that $P = A^{-1}$.

Suppose we wish to find the inverse of

$$A = \begin{bmatrix} 1 & 2 \\ 3 & 3 \end{bmatrix};$$

---

[2] H. G. Campbell, *An Introduction to Matrices, Vectors and Linear Programming* (New York, Appleton-Century-Crofts, 1965).

then

$$[A \mid I] = \begin{bmatrix} 1 & 2 & \vdots & 1 & 0 \\ 3 & 3 & \vdots & 0 & 1 \end{bmatrix} \underset{-3R_1+R_2}{\sim} \begin{bmatrix} 1 & 2 & \vdots & 1 & 0 \\ 0 & -3 & \vdots & -3 & 1 \end{bmatrix}$$

$$\underset{-\frac{1}{3}R_2}{\sim} \begin{bmatrix} 1 & 2 & \vdots & 1 & 0 \\ 0 & 1 & \vdots & 1 & -\frac{1}{3} \end{bmatrix} \underset{-2R_2+R_1}{\sim} \begin{bmatrix} 1 & 0 & \vdots & -1 & \frac{2}{3} \\ 0 & 1 & \vdots & 1 & -\frac{1}{3} \end{bmatrix}$$

$$= [I \mid A^{-1}].$$

Therefore,

$$A^{-1} = \begin{bmatrix} -1 & \frac{2}{3} \\ 1 & -\frac{1}{3} \end{bmatrix}.$$

A justification of this procedure can be given as follows: Since a non-singular matrix A can be reduced to normal form by a sequence of elementary row operations (Exercise 14), we say that there exists a matrix P (which is a product of elementary matrices) such that $PA = I$, and if we postmultiply both sides by A^{-1} we obtain

$$(PA)A^{-1} = IA^{-1},$$
$$P(AA^{-1}) = A^{-1},$$
$$PI = A^{-1},$$
$$P = A^{-1}.$$

Thus we have

$$[A \mid I] \sim P[A \mid I] = [PA \mid PI]$$
$$= [I \mid P] = [I \mid A^{-1}].$$

★ **Example 5.** The Gauss elimination method may be modified by diagonalizing, rather than triangularizing, the coefficient matrix as illustrated below; this modification is known as the *Gauss–Jordan method*. At the beginning of this section, the augmented matrix of the system

$$\begin{cases} x_2 + x_3 = 1, \\ x_2 - x_3 = 5, \\ x_1 + x_2 + x_3 = 2, \end{cases}$$

was triangularized to

$$\begin{bmatrix} 1 & 1 & 1 & \vdots & 2 \\ 0 & 1 & -1 & \vdots & 5 \\ 0 & 0 & 1 & \vdots & -2 \end{bmatrix}.$$

If we continue to apply elementary operations in order to diagonalize the coefficient matrix, we find that the augmented matrix is equivalent to

$$\begin{bmatrix} 1 & 0 & 0 & \vdots & 1 \\ 0 & 1 & 0 & \vdots & 3 \\ 0 & 0 & 1 & \vdots & -2 \end{bmatrix}.$$

The corresponding system gives $x_1 = 1$, $x_2 = 3$, $x_3 = -2$, as the obvious solution of the original system.

By now the reader should have a sufficient knowledge of matrix algebra to enable him to study some very useful applications listed in Chapters 10 and 11 of B. Noble, *Applications of Undergraduate Mathematics in Engineering* (New York, The Macmillan Company, 1967). Particular attention should be given to the applied problems presented in Section 11.2 on Gaussian elimination and the part of Section 2.6 on linear programming.

EXERCISES

1. Solve the system

$$\begin{cases} 4x + 3y = 2, \\ x + 2y = 3, \end{cases}$$

by the old addition-subtraction method. (Keep two equations at all times.) Then use the corresponding elementary row operations on the augmented matrix to achieve the same solution.

2. Which elementary row operation (if any) transforms the first matrix into the second?

$(a)\ \begin{bmatrix} 2 & 1 \\ 4 & 6 \end{bmatrix},\ \begin{bmatrix} 2 & 1 \\ 0 & 4 \end{bmatrix};$ $(b)\ \begin{bmatrix} 2 & 4 & 6 \\ 1 & 2 & 4 \end{bmatrix},\ \begin{bmatrix} 1 & 2 & 3 \\ 1 & 2 & 4 \end{bmatrix};$

$(c)\ \begin{bmatrix} 2 & 4 & 3 & 1 \\ 1 & 2 & 3 & 4 \\ 0 & 1 & 4 & 6 \end{bmatrix},\ \begin{bmatrix} 1 & 2 & 3 & 4 \\ 0 & 1 & 4 & 6 \\ 2 & 4 & 3 & 1 \end{bmatrix}.$

3. Is $\begin{bmatrix} 2 & 1 & 4 \\ 0 & 1 & 3 \\ 2 & 0 & 1 \end{bmatrix}$ equivalent to I_3? Why?

4. What is the normal form of a 4 by 6 matrix of rank 2? Express the answer using submatrices.

5. Find the rank of each of the following matrices by reducing each to normal form:

$(a)\ \begin{bmatrix} 3 & 2 & -1 \\ 7 & 8 & 0 \\ 4 & 6 & 1 \end{bmatrix};$ $(b)\ \begin{bmatrix} 3 & 4 & 0 & 6 \\ 2 & -3 & 1 & 1 \\ 7 & -2 & 2 & 8 \end{bmatrix};$ $(c)\ \begin{bmatrix} 4 & 9 & -3 & 1 \\ 6 & 9 & -4 & 0 \\ 2 & 9 & -2 & 2 \\ -2 & 0 & 1 & 1 \end{bmatrix}.$

6. Is the converse of Theorem 5.3 true? If it is not, give a counterexample.

7. In each part of this exercise, write the elementary matrix E that performs the indicated elementary transformation on

$$A = \begin{bmatrix} -3 & 2 & -1 \\ 4 & 0 & 1 \end{bmatrix}.$$

Then multiply the matrices to obtain the desired transformation.

(a) Interchange the first and second rows.
(b) Multiply the third column by 7.
(c) Add 4 times the first row to the second row.
(d) Add 5 times the third column to the first column.

8. For each of the following systems determine by the echelon method whether the system is consistent with a unique solution, and if it is, find that solution. Check your answers.

(a) $\begin{cases} x_1 + x_2 = 3, \\ 2x_1 - x_2 = 4. \end{cases}$
  $\qquad$
(b) $\begin{cases} x_1 \qquad - x_3 - 2 = 0, \\ \qquad x_2 + 3x_3 - 1 = 0, \\ x_1 - 2x_2 \qquad - 7 = 0. \end{cases}$

(c) $\begin{cases} 6x_1 + 3x_2 + 2x_3 = 1, \\ 3x_1 \qquad - 4x_3 = 4, \\ 5x_1 - x_2 \qquad = 14. \end{cases}$
  $\qquad$
(d) $\begin{cases} 2x_1 + x_2 - x_3 = -4, \\ x_1 - x_2 + 3x_3 = 3, \\ x_1 + 2x_2 - 4x_3 = 1. \end{cases}$

(e) $\begin{cases} x_1 - x_2 - x_3 - x_4 = 5, \\ x_1 + 2x_2 + 3x_3 + x_4 = -2, \\ 2x_1 \qquad + 2x_3 + 3x_4 = 3, \\ 3x_1 + x_2 \qquad + 2x_4 = 1. \end{cases}$

(f) $\begin{cases} x_1 + 2x_2 + x_3 \qquad = 2, \\ 2x_1 \qquad - 2x_3 + x_4 = 6, \\ 4x_2 + 3x_3 + 2x_4 = -1, \\ -x_1 + 6x_2 - x_3 - x_4 = 2. \end{cases}$

(g) $\begin{cases} x_1 - x_2 = 1, \\ 2x_1 + x_2 = \frac{19}{2}, \\ 4x_1 - 2x_2 = 9. \end{cases}$
Also draw a graph.
  $\qquad$
(h) $\begin{cases} x + 2y = 0, \\ 3x - y = 0. \end{cases}$

(i) $\begin{cases} x_1 + 2x_2 + 3x_3 = 0, \\ 2x_1 - x_2 + x_3 = 0, \\ 2x_1 + 3x_2 + 4x_3 = 0. \end{cases}$
  $\qquad$
(j) $\begin{cases} x_1 + x_2 + x_3 = 0, \\ x_1 - x_2 - 2x_3 = 0, \\ 2x_1 \qquad - x_3 = 0. \end{cases}$

9. Find A^{-1} by the method given in Example 4 of this section. Check by showing that $AA^{-1} = I$.

$$(a)\ A = \begin{bmatrix} 4 & -1 \\ 7 & 2 \end{bmatrix}; \quad (b)\ A = \begin{bmatrix} 1 & 0 & 1 \\ 3 & 1 & 0 \\ 1 & 0 & 0 \end{bmatrix}; \quad (c)\ A = \begin{bmatrix} 4 & 1 & 4 \\ 0 & 1 & 0 \\ 2 & 0 & 3 \end{bmatrix}.$$

10. Explain what would happen if one tried to apply the method of finding A^{-1} taught in this section to a singular matrix A.

11. Two alloys P and Q are manufactured by a certain factory, which sells its products by the ton but purchases some of the ingredients in 100-pound units. For each ton of P produced, 4 units of metal A and 4 units of metal B are required. Each ton of Q requires 7 units of A and 3 units of B. If the factory can obtain and wishes to use exactly 60 units of metal A and 40 units of metal B per day, how many tons of each alloy can it produce in a day?

12. A trucking company owns three types of trucks, numbered 1, 2, and 3, which are equipped to haul three different types of machines per load according to the following chart.

	Trucks		
	No. 1	No. 2	No. 3
Machine A	1	1	1
Machine B	0	1	2
Machine C	2	1	1

How many trucks of each type should be sent to haul exactly 12 of the type A machines, 10 of the type B machines, and 16 of the type C machines? Assume each truck is fully loaded.

13. To control a certain crop disease, it is determined that it is necessary to use 6 units of chemical A, 10 units of chemical B, and 8 units of chemical C. One barrel of commercial spray P contains 1, 3, and 4 units, respectively, of these chemicals. One barrel of commercial spray Q contains 3, 3, and 3 units, respectively, and spray R contains 2 units of A and 5 units of B. How much of each type of spray should be used to spread the exact amounts of chemicals needed to control the disease?

14. Prove that any nonzero matrix A can be reduced to normal form by elementary operations. Also, prove that if A is nonsingular, only elementary row operations are needed.

NEW VOCABULARY

§5.1	inverse matrix	§5.2	elementary row operations
§5.1	singular matrix		
§5.1	nonsingular matrix	§5.2	elementary column operations
§5.1	cofactor matrix		
§5.1	adjoint matrix	§5.2	elementary operation
§5.1	unique inverse	§5.2	row equivalent matrices
§5.2	equivalent systems of equations	§5.2	column equivalent matrices
§5.2	Gauss elimination method	§5.2	equivalent
		§5.2	normal form
§5.2	echelon method	§5.2	elementary matrix
§5.2	triangularization	§5.2	Gauss–Jordan method

Systems With More
Than One Solution

6.1 GAUSS–JORDAN ELIMINATION METHOD

Two or more planes in three-dimensional space intersecting along a line offer a geometric illustration of a system of linear equations with an infinite number of solutions.

In Chapter 4 we learned that a consistent system with more than one solution is identified by applying Theorems 4.1 and 4.2 (for such systems r exists and $r < n$). The system may then be solved by expressing certain r unknowns in terms of the other $n - r$ unknowns. (Theorem A.9, page 160.) These $n - r$ unknowns are considered to be arbitrary parameters. The solution thus obtained is called a *complete solution*. The Gauss–Jordan method presented in the last chapter will be adapted to systems with more than one solution and thus will determine complete solutions. The following two examples illustrate this very important procedure.

Example 1. Solve

$$\begin{cases} x_1 + x_2 + x_3 + x_4 = 4, \\ x_1 + 3x_2 + 3x_3 = 2, \\ x_1 + x_2 + 2x_3 - x_4 = 6. \end{cases}$$

The object of the Gauss–Jordan elimination method is to transform the augmented matrix $[A \mid B]$ of the original system, by means of elementary row operations only, to a form in which I is an r by r submatrix of the transformed coefficient matrix.

$$\begin{bmatrix} 1 & 1 & 1 & 1 & \vdots & 4 \\ 1 & 3 & 3 & 0 & \vdots & 2 \\ 1 & 1 & 2 & -1 & \vdots & 6 \end{bmatrix} \underset{-R_1+R_3}{\overset{-R_1+R_2}{\sim}} \begin{bmatrix} 1 & 1 & 1 & 1 & \vdots & 4 \\ 0 & 2 & 2 & -1 & \vdots & -2 \\ 0 & 0 & 1 & -2 & \vdots & 2 \end{bmatrix}$$

$$\underset{\frac{1}{2}R_2}{\sim} \begin{bmatrix} 1 & 1 & 1 & 1 & | & 4 \\ 0 & 1 & 1 & -\frac{1}{2} & | & -1 \\ 0 & 0 & 1 & -2 & | & 2 \end{bmatrix} \underset{-R_2+R_1}{\sim} \begin{bmatrix} 1 & 0 & 0 & \frac{3}{2} & | & 5 \\ 0 & 1 & 1 & -\frac{1}{2} & | & -1 \\ 0 & 0 & 1 & -2 & | & 2 \end{bmatrix}$$

$$\underset{-R_3+R_2}{\sim} \begin{bmatrix} 1 & 0 & 0 & \frac{3}{2} & | & 5 \\ 0 & 1 & 0 & \frac{3}{2} & | & -3 \\ 0 & 0 & 1 & -2 & | & 2 \end{bmatrix}.$$

The last matrix is the augmented matrix of the system

$$\begin{cases} x_1 & + \frac{3}{2}x_4 = & 5, \\ x_2 & + \frac{3}{2}x_4 = & -3, \\ x_3 - 2x_4 = & 2, \end{cases}$$

from which a complete solution is seen to be

$$\begin{cases} x_1 = & 5 - \frac{3}{2}x_4, \\ x_2 = & -3 - \frac{3}{2}x_4, \\ x_3 = & 2 + 2x_4. \end{cases}$$

Notice that the 3 by 3 submatrix I consists of the first three columns of the transformed matrix of coefficients.

From this complete solution we may obtain what is called a *particular solution* by assigning a value to the parameter x_4. For example, if we let $x_4 = 0$, a particular solution is $X = (5, -3, 2, 0)$; if we let $x_4 = 4$, then the vector $(-1, -9, 10, 4)$ is another particular solution. If the parameter can take on an infinite number of values, then there are an infinite number of solutions.

Example 2.   Solve

$$\begin{cases} x_1 + 2x_2 + x_3 + 5x_4 = & 3, \\ x_1 + 2x_2 + 2x_3 + 7x_4 = & 4, \\ x_3 + 2x_4 = & 1, \\ -2x_1 - 4x_2 - 6x_4 = & -4. \end{cases}$$

Using elementary row operations only, we transform the augmented matrix $[A \mid B]$ to a form in which I is an r by r submatrix of the transformed matrix of coefficients.

$$\begin{bmatrix} 1 & 2 & 1 & 5 & | & 3 \\ 1 & 2 & 2 & 7 & | & 4 \\ 0 & 0 & 1 & 2 & | & 1 \\ -2 & -4 & 0 & -6 & | & -4 \end{bmatrix} \underset{\substack{-R_1+R_2 \\ 2R_1+R_4}}{\sim} \begin{bmatrix} 1 & 2 & 1 & 5 & | & 3 \\ 0 & 0 & 1 & 2 & | & 1 \\ 0 & 0 & 1 & 2 & | & 1 \\ 0 & 0 & 2 & 4 & | & 2 \end{bmatrix}$$

$$\underset{\substack{-R_2+R_3 \\ -2R_2+R_4 \\ -R_2+R_1}}{\sim} \begin{bmatrix} 1 & 2 & 0 & 3 & | & 2 \\ 0 & 0 & 1 & 2 & | & 1 \\ 0 & 0 & 0 & 0 & | & 0 \\ 0 & 0 & 0 & 0 & | & 0 \end{bmatrix}.$$

From the last matrix we recognize that the rank is 2 and I is the 2 by 2 sub-matrix formed by deleting the last two rows and the 2nd, 4th, and 5th columns. We have the augmented matrix of the system

$$\begin{cases} x_1 + 2x_2 \quad + 3x_4 = 2, \\ \qquad\qquad x_3 + 2x_4 = 1, \end{cases}$$

from which a complete solution is easily seen to be

$$\begin{cases} x_1 = 2 - 3x_4 - 2x_2, \\ x_3 = 1 - 2x_4. \end{cases}$$

Here there are 2 parameters x_2 and x_4, and if we assign values to them, particular solutions result. For example, if we let $x_2 = 1$ and $x_4 = 2$, then the vector $(-6, 1, -3, 2)$ is a particular solution.

The reader should observe the generality and power of the Gauss–Jordan method: It may be employed on any number of linear equations involving any number of unknowns; it has the advantage of answering questions concerning the nature of a solution (Theorems 4.1 and 4.2), and at the same time it proceeds toward a solution (if there are solutions).

The Gauss–Jordan method is readily adaptable to the modern high-speed computer, thus enabling the scientist to avoid tedious calculations involved with large systems.

APPLICATIONS

★ **Example 3.**   One of the most important applications of the material of this section occurs in connection with characteristic value problems which are discussed in Chapter 8. There we seek nontrivial solutions of the homogeneous system $(A - \lambda I)X = 0$.

The reader should observe that because homogeneous systems $HX = 0$ are always consistent (since the rank of H must equal the rank of $[H \mid 0]$) and because the *trivial solution* $(0, 0, 0, \ldots, 0)$ is always a solution, then homogeneous systems possess nontrivial solutions only if there is more than one solution; the method developed in this section provides such solutions.

Example 4.   In Markov chain problems it is often desirable to determine certain limiting probabilities; for instance, in Example 8 of Section 2.4, page 34, what is the long range probability for Democratic votes, for Republican votes, and for third party votes? That is, in the long run over many elections, what percentage of the votes will be cast for the Democrats, the Republicans, and a third party? It can be shown that the solution is the probability vector satisfying the matrix equation

$$[x \quad y \quad z]P = [x \quad y \quad z],$$

where P is a regular stochastic matrix; that is, a matrix where the sum of the entries of each row is 1 and all the entries of some power P^n are positive. In the example referred to, this equation is

$$[x \quad y \quad z] \begin{bmatrix} 0.60 & 0.20 & 0.20 \\ 0.30 & 0.60 & 0.10 \\ 0.30 & 0.20 & 0.50 \end{bmatrix} = [x \quad y \quad z],$$

which amounts to solving the system

$$\begin{cases} -0.40x + 0.30y + 0.30z = 0, \\ 0.20x - 0.40y + 0.20z = 0, \\ 0.20x + 0.10y - 0.50z = 0. \end{cases}$$

It can be proved that if P is a regular stochastic matrix, the system has an infinite number of solutions. We must then choose the particular solution which is a probability vector; that is, the solution consisting of nonnegative components whose sum is equal to 1, so that the entries can be considered as probabilities. The Gauss–Jordan procedure yields the complete solution

$$x = \tfrac{9}{5}z, \qquad \text{and} \qquad y = \tfrac{7}{5}z,$$

and a particular solution whose nonnegative components add up to 1 is $(\tfrac{9}{21}, \tfrac{7}{21}, \tfrac{5}{21})$.

EXERCISES

For Exercises 1–9 use the Gauss–Jordan elimination method to determine the nature of the solution and, if more than one solution exists, find a complete solution and one particular solution.

1. $\begin{cases} x_1 + x_2 - x_3 = 4, \\ x_1 - x_2 + x_3 = 2. \end{cases}$

2. $\begin{cases} 2x_1 + x_2 - x_3 = 3, \\ x_1 + x_2 = 2, \\ x_1 - x_3 = 1. \end{cases}$

3. $\begin{cases} 3x_1 - x_2 + x_3 = 4, \\ x_1 - x_2 = 0, \\ 2x_1 + x_3 = 4, \\ 4x_1 - 2x_2 + x_3 = 4. \end{cases}$

4. $\begin{cases} x_1 + 2x_2 - x_3 + x_4 = 1, \\ x_1 - x_2 + x_3 - x_4 = 2. \end{cases}$

5. $\begin{cases} x_1 - 2x_2 + x_3 - x_4 = 3, \\ 2x_1 - 3x_2 = 3, \\ x_1 - x_2 - x_3 + x_4 = 0. \end{cases}$

6. $\begin{cases} x_1 + x_3 = 2, \\ x_2 + x_3 = 6, \\ x_2 + x_4 = 0, \\ x_1 + x_2 + x_3 + x_4 = 2. \end{cases}$

7. $\{2x_1 + x_2 = 4.$

8. $\begin{cases} 2x_1 - x_2 = 0, \\ -4x_1 + 2x_2 = 0. \end{cases}$

9. $\begin{cases} x_1 - 3x_2 + 2x_3 = 0, \\ -x_1 - 2x_2 + 2x_3 = 0, \\ -2x_1 + x_2 = 0. \end{cases}$

10. Explain how the Gauss–Jordan method determines whether or not a system is inconsistent.

11. State a necessary and sufficient condition for a homogeneous system to have a nontrivial solution.

12. Demonstrate graphically that the system

$$\begin{cases} x + z = 1, \\ x + y + z = 3, \\ y = 2, \end{cases}$$

has an infinite number of solutions. Then, using the Gauss–Jordan method, find the complete solution and three particular solutions (be sure that the particular solutions agree with the graph).

13. (a) A trucking company owns three types of trucks which are equipped to haul two different types of machines per load according to the following chart:

	Trucks		
	No. 1	No. 2	No. 3
Machine A	1	1	1
Machine C	2	1	1

How many trucks of each type could be sent to haul exactly 12 of the type A machine and 16 of the type C machines? Assume each truck is fully loaded. (See Exercise 12 of Section 5.2.)

(b) Why are there not an infinite number of solutions to part (a)?

14. To control a certain crop disease, it is determined that it is necessary to use 6 units of chemical A and 10 units of chemical B. One barrel of commercial spray P contains 1 unit of A and 3 units of B. One barrel of spray Q

contains 3 units of A and 3 units of B, and spray R contains 2 units of A and 5 units of B. How much of each type of spray could be used to spread the exact amount of chemicals needed to control the disease? (See Exercise 13 of Section 5.2).

6.2 BASIC VARIABLES (*OPTIONAL*)

Example 1. A complete solution of the system

$$\begin{cases} x_1 + x_2 + 3x_3 = 8, \\ x_2 + 3x_3 = 6, \\ x_1 + 2x_2 + 6x_3 = 14, \end{cases}$$

can be determined to be

$$\begin{cases} x_1 = 2, \\ x_2 = 6 - 3x_3. \end{cases}$$

Here, x_3 is the parameter. The question then arises, are there other complete solutions? That is, can x_2 or x_1 serve as parameters? In this simple problem it is easy to observe that we can solve for x_1 and x_2 in terms of x_3, and we can solve for x_1 and x_3 in terms of x_2, but we *cannot* solve for x_2 and x_3 in terms of x_1.

In problems involving many unknowns it is sometimes advantageous to be able to predict which unknowns can serve as parameters and which cannot without actually finding the solution(s). This is a motivation behind Theorem 6.1.

Theorem 6.1. *A consistent system of linear equations of rank r can be solved for r unknowns, say $x_{i_1}, x_{i_2}, \ldots, x_{i_r}$ in terms of the remaining $n - r$ unknowns, if and only if the submatrix of coefficients of $x_{i_1}, x_{i_2}, \ldots, x_{i_r}$ has rank r.*

Proof. See Theorem A.9, page 160.

In Example 1, the rank of the system is 2. The rank of the coefficient matrix for x_1 and x_2 is 2 since

$$\begin{bmatrix} 1 & 1 \\ 0 & 1 \\ 1 & 2 \end{bmatrix} \sim \begin{bmatrix} 1 & 0 \\ 0 & 1 \\ 0 & 0 \end{bmatrix}.$$

This allows us to predict the solution of x_1 and x_2 in terms of x_3. However, the rank of the matrix of coefficients for x_2 and x_3 is 1 because

$$\begin{bmatrix} 1 & 3 \\ 1 & 3 \\ 2 & 6 \end{bmatrix} \sim \begin{bmatrix} 1 & 0 \\ 0 & 0 \\ 0 & 0 \end{bmatrix}.$$

This allows us to predict that x_2 and x_3 cannot be solved in terms of x_1.

In Theorem 6.1 if there are m equations where $m < n$ and $r = m$ then we have a special class of problems which are very important. Consider the complete solution of such a problem: If we assign the value zero to each of the parameters, the resulting particular solution is called a **basic solution.** The m unknowns, or variables, not serving as parameters in a basic solution are called **basic variables.**

Example 2.   The system

$$\begin{cases} x_1 + x_2 + 3x_3 - x_4 = 6, \\ x_1 + 2x_2 + 2x_3 - x_4 = 2, \end{cases}$$

can be shown to have a complete solution

$$\begin{cases} x_1 \quad = 10 - 4x_3 + x_4, \\ \quad x_2 = -4 + x_3. \end{cases}$$

When the parameters x_3 and x_4 are both zero, there results a basic solution $(10, -4, 0, 0)$. Here the basic variables are x_1 and x_2. This, of course, is not the only basic solution. For a complete solution in which another pair of unknowns are the parameters, there would be a different basic solution.

APPLICATIONS

The concepts introduced in this section form the basis of a relatively new branch of applied mathematics known as linear programming. Scholars in managerial science, agriculture, economics, and some areas of engineering find linear programming to be very useful. The following example gives the reader a small insight into the nature of linear programming and its relation to the material of this chapter.

Example 3. Consider a manufacturer who can produce two types of products M and N on machine A and the same two products on machine B according to the following chart.

	Product M	Product N	Machine limitation
Machine A	1 hour per unit	4 hours per unit	16 hours per day
Machine B	3 hours per unit	2 hours per unit	18 hours per day
Profit per unit	$ 200	$ 300	

The profit for the production of one unit of each product and the machine limitations per day are also given in the chart. The manufacturer would like to know how many units of each product he should manufacture per day in order to stay within the limitations of his machines and still be able to maximize his profit. Suppose we let x_1 represent the number of units of M he produces per day and let x_2 represent the number of units of N he produces per day; then the problem becomes maximize P where

$$P = 200x_1 + 300x_2,$$

subject to the conditions

$$\begin{cases} x_1 + 4x_2 \leq 16, \\ 3x_1 + 2x_2 \leq 18, \\ x_1 \geq 0, \\ x_2 \geq 0. \end{cases}$$

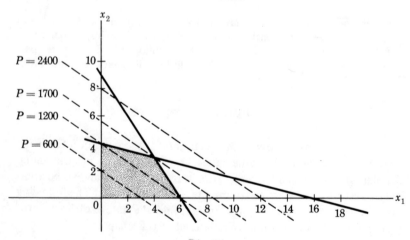

Fig. 6.1

A graphical solution to this problem is given in Figure 6.1. The shaded region of the graph is the set of points which satisfy all of the inequalities simultaneously; such a set is known as a *feasible set* (See Example 8 of Section 4.1.) With P as a parameter, the graph of $P = 200x_1 + 300x_2$ is a family of parallel lines; the lines correspond to positive increasing values of the profit P. In Figure 6.1 it appears that the line corresponding to $P = 1,700$ is the "highest" line that has a point in common with the shaded region. Any higher line does not intersect the feasible set and lower lines represent smaller values of P. Hence the point (4, 3) yields:

$$\text{maximum } P = (200)(4) + (300)(3) = 1700.$$

The reader should observe that there are four "corners" (called *extreme points*) of the shaded region and that if a solution exists, it occurs at an extreme point.

Obviously, if the number of manufactured products is greater than 3, a geometrical method of solution would be impossible, hence the need for the following algebraic approach. To solve this example algebraically we change the first two inequalities to equalities by introducing two more nonnegative variables x_3 and x_4 so that our problem now becomes maximize P where

$$P = 200x_1 + 300x_2 + 0x_3 + 0x_4,$$

subject to the conditions

$$\begin{cases} x_1 + 4x_2 + x_3 & = 16, \\ 3x_1 + 2x_2 + x_4 & = 18, \\ x_1 & \geq 0, \\ x_2 & \geq 0, \\ x_3 & \geq 0, \\ x_4 & \geq 0. \end{cases}$$

An important theorem of linear programming states that if P_{max} exists, it occurs for a basic solution in which $X \geq 0$. Because there are always a finite number of basic solutions, the problem can be resolved. In this problem the basic solutions are

$$(0, 0, 16, 18), \qquad (6, 0, 10, 0), \qquad (0, 4, 0, 10),$$
$$(16, 0, 0, -30), \qquad (0, 9, -20, 0), \qquad (4, 3, 0, 0).$$

Note that the first two coordinates of each basic solution correspond to the intersections of each pair of lines in the graph and further notice that there is a one-to-one correspondence between the extreme points of the feasible set and the nonnegative basic solutions. If we substitute each nonnegative basic solution into

$$P = 200x_1 + 300x_2 + 0x_3 + 0x_4$$

we find that (4, 3) yields maximum P.

EXERCISES

1. In the following consistent system, can we solve for x_1 and x_2 in terms of x_3? Why?

$$\begin{cases} x_1 + x_2 + x_3 = 4, \\ -x_1 - x_2 \qquad = 2, \\ 2x_1 + 2x_2 + 2x_3 = 8. \end{cases}$$

2. Without solving the system, state which unknowns can be expressed in terms of the other unknowns in the following system:

$$\begin{cases} x_1 - x_2 - 3x_3 - 2x_4 = 4, \\ x_1 + x_2 - 3x_3 - 2x_4 = 2. \end{cases}$$

State reasons for your conclusions.

3. (a) Determine one basic solution for

$$\begin{cases} -x_1 + x_2 + x_3 \qquad = 4, \\ 2x_1 - 2x_2 \qquad + x_4 = 5. \end{cases}$$

(b) Determine which sets of unknowns may serve as basic variables.

4. In the following system, the ranks of the coefficient and augmented matrices are 2. Solve for x and y in terms of z and w.

$$\begin{cases} x + y + z + 2w = 1, \\ 2x + 2y + 2z + 4w = 2, \\ 2x - y + 2z + w = 2. \end{cases}$$

5. Express the following systems of inequalities as systems of equations:

(a) $\begin{cases} z_1 + 2z_2 \le 5, \\ z_1 + 3z_2 \le 7. \end{cases}$

(b) $\begin{cases} z_1 + z_2 \le 4, \\ z_1 + 4z_2 \le 7, \\ 2z_2 \le 3. \end{cases}$

6. (a) Represent graphically the system of inequalities in Exercise 5(a) with the additional restriction $Z \ge 0$.

(b) The resulting feasible set has 4 bounding lines; each line intersects every other line at a particular point and there are 6 such intersection points. Find these points and state which are the extreme points.

(c) The system of equations, consisting of the answer to Exercise 5(a), has 6 basic solutions. Find these basic solutions and show how they correspond to the 6 points found in part (b) of this problem. How can one distinguish which of the basic solutions represent extreme points?

7. A local television network is faced with the following problem. It has been found that program A with 20 minutes of music and 1 minute of advertisement draws 30,000 viewers while program B with 10 minutes of music and 1 minute of advertisement draws 10,000 viewers. The advertiser insists that at least 6 minutes per week be devoted to his advertisement and the network can afford no more than 80 minutes of music per week. How many times per week should each program be given in order to obtain the maximum number of viewers?

8. State the mathematical formulation of the linear programming problem in Example 3 using matrix notation.

9. In Example 8 of Section 4.1, suppose that both brands of fertilizer yield a profit of 1¢ per pound. How much of each brand should the manufacturer produce in order to maximize his profit?

NEW VOCABULARY

§6.1	complete solution	§6.2	basic solution
§6.1	particular solution	§6.2	basic variables
§6.1	trivial solution	§6.2	extreme points

Part III:

Matrix Transformations

Linear Transformations

7.1 GEOMETRIC INTERPRETATION OF A MAPPING

A unary operation transforms the elements of one set into unique elements of another set and is a very important concept in mathematics. Such an operation frequently is called a **transformation** or a *function* or a **mapping.** For example, the scalar x can be mapped into the scalar x^3. Let this transformation be represented by T; it can be expressed in words as "the cube of," and denoted in the following ways

$$T(x) = x^3 \qquad \text{or} \qquad x \xrightarrow{T} x^3.$$

The element x^3 is called the *image* of x. In Chapter 3 we studied the mapping

$$T(A) = \det A \qquad \text{or} \qquad A \xrightarrow{T} \det A.$$

For a mapping of the elements of a set U into unique elements of a set V, the set U is called the **domain** of the mapping. The subset of elements of V, each of which is actually an image of at least one element of U, is called the **range** of the mapping.

Another transformation "the magnitude of" maps a real vector α into a scalar $|\alpha|$ or in mathematical notation

$$T(\alpha) = |\alpha| \qquad \text{or} \qquad \alpha \xrightarrow{T} |\alpha|.$$

Here, the domain is the set of n-dimensional real vectors, and the range is the set of nonnegative real numbers (see Example 4 of Section 1.3). Still other transformations map vectors into vectors. We devote the rest of this section to a geometric discussion of such mappings (in two dimensions) in order to illustrate a variety of vector transformations. All vectors used in the following examples originate at the origin. The following transformations may be considered as vector functions of a vector variable.

Example 1. Let T map each nonzero vector α into the vector which has the same direction but twice the magnitude of α, and let T map the zero vector onto itself. The transformation T is illustrated in Figure 7.1. Such a transformation may be referred to as a "stretching."

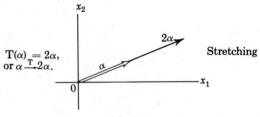

Fig. 7.1

Example 2. Let T map each nonzero vector α into the vector having the same magnitude but the opposite direction, and let T map the zero vector onto itself. This transformation is shown in Figure 7.2. Such a transformation is referred to as a "reflection through the origin."

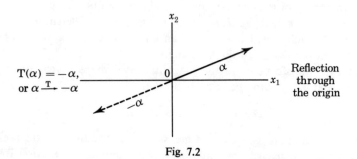

Fig. 7.2

Example 3. Let T be the mapping which reflects each vector $\alpha = (a_1, a_2)$ through the x_1-axis ($a_2 \neq 0$), and let T map α onto itself if $a_2 = 0$. (Figure 7.3.).

Example 4. Let T be a mapping which gives a 90° rotation in a counterclockwise direction with respect to the origin to each vector $\alpha = (a_1, a_2)$, and let T map the zero vector onto itself. (Figure 7.4.)

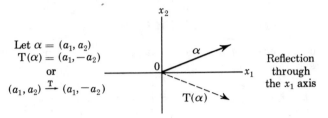

Let $\alpha = (a_1, a_2)$
$T(\alpha) = (a_1, -a_2)$
or
$(a_1, a_2) \xrightarrow{\ T\ } (a_1, -a_2)$

Reflection
through
the x_1 axis

Fig. 7.3

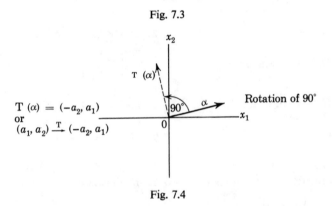

$T(\alpha) = (-a_2, a_1)$
or
$(a_1, a_2) \xrightarrow{\ T\ } (-a_2, a_1)$

Rotation of 90°

Fig. 7.4

Example 5. Let T be a mapping which slides the terminal point of a vector $\alpha = (a_1, a_2)$ parallel to the x_1-axis a distance of $3a_2$ units with the initial point remaining fixed at the origin ($a_2 \neq 0$), and let T map α onto itself if $a_2 = 0$. (Figure 7.5). Such a transformation may be referred to as a "shear."

Applications of the material of this chapter are deferred until the next chapter.

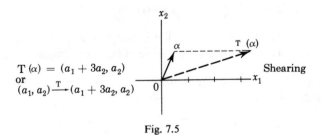

$T(\alpha) = (a_1 + 3a_2, a_2)$
or
$(a_1, a_2) \xrightarrow{\ T\ } (a_1 + 3a_2, a_2)$

Shearing

Fig. 7.5

EXERCISES

1. Find the image of β under the mapping $\beta \xrightarrow{T} -2\beta$, or $T(\beta) = -2\beta$. Then graph β and the image of β.

(a) Let $\beta = \begin{bmatrix} 1 \\ -2 \end{bmatrix}$;   (b) let $\beta = \begin{bmatrix} 0 \\ 3 \end{bmatrix}$;   (c) let $\beta = -2\begin{bmatrix} 1 \\ 1 \end{bmatrix}$.

In Exercises 2–9, what are the domain and range for the following mappings? Let x and y represent real numbers.

2. $T(x) = x^2 + 1$.

3. $x \xrightarrow{T} (x, 2x)$.

4. $T(\alpha) = \alpha^2$, where $\alpha = (x, y)$.

5. $\begin{bmatrix} x \\ y \end{bmatrix} \xrightarrow{T} \begin{bmatrix} 0 \\ y \end{bmatrix}$.

6. $T(A) = A^2$, where $A = \begin{bmatrix} x & x \\ x & x \end{bmatrix}$.

7. $T(x) = \sqrt{1 - x^2}$.

8. $T(x) = -\sqrt{1 - x^2}$.

9. $T(x) = \sqrt{x^2 - 1} + 1$.

In Exercises 10–15, describe the geometric effect of each of the following transformations. State this effect in words and illustrate with a graph.

10. $\begin{bmatrix} x_1 \\ x_2 \end{bmatrix} \xrightarrow{T} \begin{bmatrix} 2x_1 \\ x_2 \end{bmatrix}$.

11. $T(\alpha) = -2\alpha$ where $\alpha = \begin{bmatrix} a_1 \\ a_2 \end{bmatrix}$.

12. $\begin{bmatrix} x_1 \\ x_2 \end{bmatrix} \xrightarrow{T} \begin{bmatrix} x_1 \\ 0 \end{bmatrix}$.

13. $\begin{bmatrix} x_1 \\ x_2 \end{bmatrix} \xrightarrow{T} \begin{bmatrix} x_1 + x_2 \\ x_2 \end{bmatrix}$.

14. $(x_1, x_2) \xrightarrow{T} (x_1 + 2, x_2 + 3)$.

15. $\begin{bmatrix} x_1 \\ x_2 \end{bmatrix} \xrightarrow{T} \begin{bmatrix} -2x_2 \\ 2x_1 \end{bmatrix}$.

7.2 MATRIX TRANSFORMATIONS

The various transformations illustrated in Examples 1–5 of Section 7.1 can also be expressed by the use of matrix multiplication.

Example 1. If we use a column matrix to represent α, the "stretching" transformation $\alpha \xrightarrow{T} 2\alpha$ may be expressed as

$$\begin{bmatrix} a_1 \\ a_2 \end{bmatrix} \xrightarrow{T} \begin{bmatrix} 2 & 0 \\ 0 & 2 \end{bmatrix} \begin{bmatrix} a_1 \\ a_2 \end{bmatrix} = \begin{bmatrix} 2a_1 \\ 2a_2 \end{bmatrix}.$$

Example 2. Likewise the "reflection through the origin" transformation $\alpha \overset{T}{\rightarrow} -\alpha$ may be expressed as

$$\begin{bmatrix} a_1 \\ a_2 \end{bmatrix} \overset{T}{\rightarrow} \begin{bmatrix} -1 & 0 \\ 0 & -1 \end{bmatrix} \begin{bmatrix} a_1 \\ a_2 \end{bmatrix} = \begin{bmatrix} -a_1 \\ -a_2 \end{bmatrix}.$$

Example 3. The "reflection through the x_1-axis" transformation $(a_1, a_2) \overset{T}{\rightarrow} (a_1, -a_2)$ may be expressed as

$$\begin{bmatrix} a_1 \\ a_2 \end{bmatrix} \overset{T}{\rightarrow} \begin{bmatrix} 1 & 0 \\ 0 & -1 \end{bmatrix} \begin{bmatrix} a_1 \\ a_2 \end{bmatrix} = \begin{bmatrix} a_1 \\ -a_2 \end{bmatrix}.$$

Example 4. The "90° rotation" of a nonzero vector is expressed as

$$(a_1, a_2) \overset{T}{\rightarrow} (-a_2, a_1).$$

This transformation may be written in matrix notation as

$$\begin{bmatrix} a_1 \\ a_2 \end{bmatrix} \overset{T}{\rightarrow} \begin{bmatrix} 0 & -1 \\ 1 & 0 \end{bmatrix} \begin{bmatrix} a_1 \\ a_2 \end{bmatrix} = \begin{bmatrix} -a_2 \\ a_1 \end{bmatrix}.$$

Example 5. The "shearing" transformation

$$(a_1, a_2) \overset{T}{\rightarrow} (a_1 + 3a_2, a_2)$$

may be expressed as

$$\begin{bmatrix} a_1 \\ a_2 \end{bmatrix} \overset{T}{\rightarrow} \begin{bmatrix} 1 & 3 \\ 0 & 1 \end{bmatrix} \begin{bmatrix} a_1 \\ a_2 \end{bmatrix} = \begin{bmatrix} a_1 + 3a_2 \\ a_2 \end{bmatrix}.$$

Examples 1–5, above, illustrate mappings of vectors into unique images by matrix multiplication. We formalize this concept by the following definition. A mapping of the form $\alpha \overset{T}{\rightarrow} A\alpha$ is a ***matrix transformation of a vector.***

The introduction of new variables expressed as linear combinations of the original variables is an important example of a matrix transformation. Consider

$$\begin{cases} \bar{x}_1 = a_{11}x_1 + a_{12}x_2 + \cdots + a_{1n}x_n, \\ \bar{x}_2 = a_{21}x_1 + a_{22}x_2 + \cdots + a_{2n}x_n, \\ \quad \cdot \\ \quad \cdot \\ \quad \cdot \\ \bar{x}_n = a_{n1}x_1 + a_{n2}x_2 + \cdots + a_{nn}x_n, \end{cases}$$

or

$$\bar{X} = AX \qquad \text{or} \qquad X \overset{T}{\rightarrow} AX.$$

Here we have mapped any point of the X-coordinate system into an image in the $\bar{X}$-coordinate system. The $\bar{X}$-coordinate system *need not* be considered as separate from the X-coordinate system. Actually two different interpretations of this transformation may be given; they are illustrated in the following example.

Example 6. Consider the effect of the transformation of rotation through the angle θ on the point (x_1, x_2) in the $x_1 x_2$-plane. This transformation yields the image point $(\bar{x}_1, \bar{x}_2)$ in the $\bar{x}_1 \bar{x}_2$-plane according to the equations

$$\begin{cases} \bar{x}_1 = (\cos\theta)x_1 + (-\sin\theta)x_2, \\ \bar{x}_2 = (\sin\theta)x_1 + (\cos\theta)x_2, \end{cases}$$

or

$$\begin{bmatrix} \bar{x}_1 \\ \bar{x}_2 \end{bmatrix} = \begin{bmatrix} \cos\theta & -\sin\theta \\ \sin\theta & \cos\theta \end{bmatrix} \begin{bmatrix} x_1 \\ x_2 \end{bmatrix},$$

or

$$\bar{X} = AX \qquad \text{or} \qquad X \xrightarrow{T} AX.$$

Specifically, let $\theta = 30°$, and let $(x_1, x_2) = (2, 0)$. Then

$$\begin{bmatrix} \bar{x}_1 \\ \bar{x}_2 \end{bmatrix} = \begin{bmatrix} \frac{1}{2}\sqrt{3} & -\frac{1}{2} \\ \frac{1}{2} & \frac{1}{2}\sqrt{3} \end{bmatrix} \begin{bmatrix} 2 \\ 0 \end{bmatrix} = \begin{bmatrix} \sqrt{3} \\ 1 \end{bmatrix}.$$

Thus the image of $(2, 0)$ is $(\sqrt{3}, 1)$. This transformation may be interpreted in two ways:

(1) Let the $\bar{x}_1 \bar{x}_2$-plane be the same as the $x_1 x_2$-plane and thus the point $(2, 0)$ has been mapped into another point $(\sqrt{3}, 1)$ in the same plane (Figure 7.6).

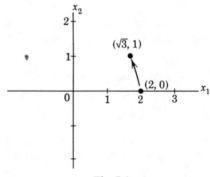

Fig. 7.6

(2) Let the point remain fixed and consider the original axes as having been rotated through $-30°$. Therefore the same point that has the coordinates $(2, 0)$ in the original x_1x_2-plane has the coordinates $(\sqrt{3}, 1)$ in the new or image $\bar{x}_1\bar{x}_2$-plane (Figure 7.7).

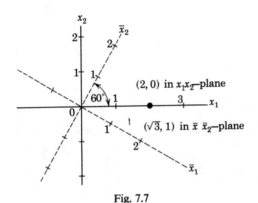

Fig. 7.7

The introduction of a single new variable into a system of linear equations is an example of a *matrix transformation of a matrix*.

Example 7. Consider the inconsistent system

$$\begin{cases} (1)\ x_1 + x_2 = 1, \\ (2)\ x_1 \qquad = 0, \\ (3)\ \qquad x_2 = 0. \end{cases}$$

The introduction of the new variable $\bar{x}_1 = x_1 + x_2 - 1$ transforms the system to

$$\begin{cases} (1')\ \bar{x}_1 \qquad = \quad 0, \\ (2')\ \bar{x}_1 - x_2 = -1, \\ (3')\ \qquad x_2 = \quad 0. \end{cases}$$

Figures 7.8 and 7.9, respectively, show the graphs before and after the transformation.

This transformation could have been accomplished by postmultiplying the original augmented matrix by a product of elementary matrices, which are designed to add (-1) times the first column to the second column and to add

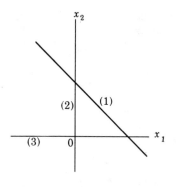

Fig. 7.8   Fig. 7.9

(−1) times the first column to the third column, to produce the final augmented matrix, or

$$
\begin{bmatrix} 1 & 1 & \vdots & 1 \\ 1 & 0 & \vdots & 0 \\ 0 & 1 & \vdots & 0 \end{bmatrix}
\begin{bmatrix} 1 & -1 & -1 \\ 0 & 1 & 0 \\ 0 & 0 & 1 \end{bmatrix}
=
\begin{bmatrix} 1 & 0 & \vdots & 0 \\ 1 & -1 & \vdots & -1 \\ 0 & 1 & \vdots & 0 \end{bmatrix}.
$$

$$
\begin{bmatrix} \text{original} \\ \text{augmented} \\ \text{matrix} \end{bmatrix}
\begin{bmatrix} \text{Product of} \\ \text{elementary} \\ \text{matrices} \end{bmatrix}
=
\begin{bmatrix} \text{final} \\ \text{augmented} \\ \text{matrix} \end{bmatrix}
$$

We have thus mapped a matrix into another matrix by matrix multiplication.

In the preceding example the x_1 and $\bar{x}_1$-axes were considered to be the
same, and the net result was the rotation and translation of two of the
three lines.

Note that in a system of linear equations, a change in variables may
be effected by elementary column operations; the resulting system is
not equivalent to the original system. On the other hand, elementary
row operations maintain the same variables and the resulting system is
equivalent to the original system.

EXERCISES

1. The mapping $\beta \to k\beta$ is the "stretching" transformation described in
Example 1 of Section 7.1. This can be thought of as a change of scale and
hence the number k is the scale factor or "scalar." Determine the matrix A
such that

$$
\beta \xrightarrow{T} A\beta = k\beta, \quad \text{where } \beta = \begin{bmatrix} b_1 \\ b_2 \end{bmatrix}.
$$

2. Determine the image of $\alpha = \begin{bmatrix} a_1 \\ a_2 \end{bmatrix}$ under the matrix transformation $\alpha \xrightarrow{T} A\alpha$ for each A, and describe the geometric effect of the transformation.

(a) $A = \begin{bmatrix} 1 & 0 \\ 0 & 1 \end{bmatrix}$; (b) $A = \begin{bmatrix} 2 & 0 \\ 0 & 1 \end{bmatrix}$; (c) $A = \begin{bmatrix} 0 & 0 \\ 0 & 1 \end{bmatrix}$;

(d) $A = \begin{bmatrix} 0 & -3 \\ 3 & 0 \end{bmatrix}$; (e) $A = \begin{bmatrix} 1 & 1 \\ 0 & 1 \end{bmatrix}$; (f) $A = \begin{bmatrix} 1 & 1 \\ 1 & 1 \end{bmatrix}$.

3. Find a matrix A such that the following transformations are accomplished by $T(\alpha) = A\alpha$.

(a) α is projected on the x_1-axis.
(b) α is reflected through the x_1-axis and the magnitude is doubled.
(c) α is rotated $45°$ in the x_1x_2-plane.

4. (a) Consider the system of equations

$$\begin{cases} x_1 + 2x_2 = 2, \\ x_1 \qquad\;\; = 0, \\ \qquad x_2 = 0, \end{cases}$$

and introduce the new variable $\bar{x}_1 = x_1 + 2x_2 - 2$. What is the new system? Graph both systems.

(b) This transformation could be accomplished by elementary column operations on the augmented matrix. These operations are (i) add -2 times the first column to the second column and (ii) add -2 times the first column to the third column. What are the two elementary matrices which accomplish these operations?

(c) Express the transformation of part (a) as $T(F) = FQ = G$, where F is the augmented matrix of the original system, G is the augmented matrix of the second system, and Q is a product of elementary matrices.

7.3 LINEAR TRANSFORMATIONS

The purpose of this section is to define a linear transformation, which is the type of transformation with which we are primarily concerned, and then to show that the transformations of Section 7.2 are of this type.

Definition 7.1. *A transformation T for which*

$$T(\alpha + \beta) = T(\alpha) + T(\beta) \qquad and \qquad T(k\alpha) = kT(\alpha),$$

where k is a scalar, is called a **linear transformation.**

We can obtain a multitude of examples of linear transformations if we can show that the matrix transformations discussed in Section 7.2 are linear. Let A be an m by n matrix and consider the matrix transformation T expressed by premultiplying a vector by A. If α and β are two vectors and k is a scalar, we have

$$A(\alpha + \beta) = A\alpha + A\beta \qquad \text{and} \qquad A(k\alpha) = k(A\alpha);$$

hence the requirements of a linear transformation are met, and thus we have the following theorem. (The reasons for these statements are left as exercises for the student.)

Theorem 7.1. *Every matrix transformation of a vector is a linear transformation.*

All of the examples of the preceding section are examples of linear transformations. The elementary row transformations of Section 5.2 are also examples of linear transformations. There the elements which were transformed as well as their images were matrices rather than vectors, however.

An example of a transformation that is nonlinear may be enlightening.

Example 1. Let the transformation T map a two-dimensional vector (a, b) into $(a + 2, b)$. This transformation is not linear because

$$T\{k(a, b)\} = T\{(ka, kb)\} = (ka + 2, kb),$$

whereas

$$kT\{(a, b)\} = k(a + 2, b) = (ka + 2k, kb),$$

and, in general, these two results are not equal. Furthermore,

$$T\{(a, b) + (c, d)\} = T\{(a + c, b + d)\} = (a + c + 2, b + d),$$

whereas

$$T\{(a, b)\} + T\{(c, d)\} = (a + 2, b) + (c + 2, d) = (a + c + 4, b + d),$$

and these two results are not equal.

EXERCISES

1. Show that the following transformation is not linear.

$$(a, b) \xrightarrow{T} (a, 2) \qquad \text{or} \qquad T(a, b) = (a, 2).$$

2. In Exercise 1, is it possible to express the transformation as

$$\begin{bmatrix} a \\ b \end{bmatrix} \xrightarrow{T} A \begin{bmatrix} a \\ b \end{bmatrix} = \begin{bmatrix} a \\ 2 \end{bmatrix},$$

where A is a 2 by 2 matrix? Why?

3. Show by definition that the following transformation is linear.

$$(a, b) \xrightarrow{T} (-2a, 2b) \qquad \text{or} \qquad T(a, b) = (-2a, 2b).$$

4. Find out which of the following transformations are linear and justify your answer.

(a) $(a, b) \xrightarrow{T} (a + 1, b)$;

(b) $\begin{bmatrix} a \\ b \end{bmatrix} \xrightarrow{T} \begin{bmatrix} a + b \\ a - b \end{bmatrix}$;

(c) $T(a, b) = (3a, 2b)$;

(d) $\begin{bmatrix} a \\ b \end{bmatrix} \xrightarrow{T} \frac{1}{2}\left(\begin{bmatrix} a \\ b \end{bmatrix} + \begin{bmatrix} 2 \\ 1 \end{bmatrix} \right)$;

(e) $T(\alpha) = |\alpha|\alpha$, where $\alpha = (a_1, a_2)$;

(f) $T(a, b) = (0, 0)$.

5. Give the reasons for the statements in the proof of Theorem 7.1, given in the paragraph preceding the Theorem.

NEW VOCABULARY

§7.1   mapping
§7.1   image
§7.1   domain
§7.1   range
§7.1   stretching

§7.1   reflection
§7.1   rotation
§7.1   shearing
§7.2   matrix transformation
         of a vector
§7.3   linear transformation

Characteristic Value Problems

8.1 CHARACTERISTIC VALUES AND VECTORS OF A MATRIX

In this section we consider a problem of matrix algebra that is met in a variety of applications. The problem is rather simply stated: For a given nth-order matrix A and for an unknown column vector

$$X = \begin{bmatrix} x_1 \\ x_2 \\ \cdot \\ \cdot \\ \cdot \\ x_n \end{bmatrix}, \text{ find values of the scalar } \lambda \text{ for which the equation}$$

$$AX = \lambda X$$

has nontrivial solutions, that is, $X \neq 0$. The problem can be restated as a homogeneous system of linear equations

$$(A - \lambda I)X = 0,$$

which we know has a solution other than $X = 0$ if and only if the rank of $A - \lambda I$ is less than n; that is, if $\det (A - \lambda I) = 0$. The equation $\det (A - \lambda I) = 0$ is known as the *characteristic equation*, and its roots λ_i are known as the *characteristic values* of the matrix A.

Example 1. For the matrix

$$A = \begin{bmatrix} 1 & 2 \\ -1 & 4 \end{bmatrix},$$

$AX = \lambda X$ is

$$\begin{bmatrix} 1 & 2 \\ -1 & 4 \end{bmatrix} \begin{bmatrix} x_1 \\ x_2 \end{bmatrix} = \lambda \begin{bmatrix} x_1 \\ x_2 \end{bmatrix},$$

or

$$\left(\begin{bmatrix} 1 & 2 \\ -1 & 4 \end{bmatrix} - \lambda \begin{bmatrix} 1 & 0 \\ 0 & 1 \end{bmatrix} \right) \begin{bmatrix} x_1 \\ x_2 \end{bmatrix} = \begin{bmatrix} 0 \\ 0 \end{bmatrix}.$$

Now a solution $X \neq 0$ exists for the above homogeneous system if and only if

$$\det\left(\begin{bmatrix} 1 & 2 \\ -1 & 4 \end{bmatrix} - \lambda \begin{bmatrix} 1 & 0 \\ 0 & 1 \end{bmatrix}\right) = 0.$$

That is,

$$\det\begin{bmatrix} 1 - \lambda & 2 \\ -1 & 4 - \lambda \end{bmatrix} = 0.$$

The expansion of this determinant yields the characteristic equation $\lambda^2 - 5\lambda + 6 = 0$, whose roots are $\lambda_1 = 2$ and $\lambda_2 = 3$. These roots are the characteristic values of the original matrix A.

Once the characteristic values have been found, we can then solve the homogeneous system

$$(A - \lambda I)X = 0$$

for X. The resulting nonzero vectors $X = X_i$ are the **characteristic vectors**. These vectors are said to be **normalized** if each one is multiplied by the reciprocal of its magnitude. In other words, a normalized vector has a magnitude of one unit.

Example 2. In the preceding example we found $\lambda_1 = 2$, $\lambda_2 = 3$. The problem now is to find the characteristic vectors that correspond to the respective characteristic values; that is, find nonzero solutions of

$$(A - 2I)X = 0 \qquad \text{and} \qquad (A - 3I)X = 0,$$

or

$$\begin{cases} -x_1 + 2x_2 = 0, \\ -x_1 + 2x_2 = 0, \end{cases} \quad \text{and} \quad \begin{cases} -2x_1 + 2x_2 = 0, \\ -x_1 + x_2 = 0. \end{cases}$$

The respective augmented matrices are

$$\begin{bmatrix} -1 & 2 & \vdots & 0 \\ -1 & 2 & \vdots & 0 \end{bmatrix} \quad \text{and} \quad \begin{bmatrix} -2 & 2 & \vdots & 0 \\ -1 & 1 & \vdots & 0 \end{bmatrix}.$$

The Gauss elimination method produces

$$\begin{bmatrix} 1 & -2 & \vdots & 0 \\ 0 & 0 & \vdots & 0 \end{bmatrix} \quad \text{and} \quad \begin{bmatrix} 1 & -1 & \vdots & 0 \\ 0 & 0 & \vdots & 0 \end{bmatrix}.$$

Therefore, the respective complete solutions are

$$x_1 = 2x_2 \qquad \text{and} \qquad x_1 = x_2.$$

Hence,

for $\lambda = 2$, $X_1 = \begin{bmatrix} x_1 \\ x_2 \end{bmatrix} = \begin{bmatrix} 2x_2 \\ x_2 \end{bmatrix}$,   and   for $\lambda = 3$, $X_2 = \begin{bmatrix} x_1 \\ x_2 \end{bmatrix} = \begin{bmatrix} x_2 \\ x_2 \end{bmatrix}$, where the x_2 of each characteristic vector is an arbitrary, nonzero, real parameter. Suppose we arbitrarily let the parameter be 1 in X_1 and -3 in X_2; we then obtain $X_1 = \begin{bmatrix} 2 \\ 1 \end{bmatrix}$ and $X_2 = \begin{bmatrix} -3 \\ -3 \end{bmatrix}$. Notice that neither X_1 nor X_2 is unique. If these characteristic vectors are normalized, we obtain

$$\begin{bmatrix} 2/\sqrt{5} \\ 1/\sqrt{5} \end{bmatrix} \quad \text{and} \quad \begin{bmatrix} -1/\sqrt{2} \\ -1/\sqrt{2} \end{bmatrix}.$$

Observe that normalized characteristic vectors are unique except for a possible change in sign of each vector.

The left-hand side, $|A - \lambda I|$, of the characteristic equation is known as the *characteristic function* or the *characteristic polynomial* of A. The following well-known theorem establishes a relationship between A and its characteristic function.

Theorem 8.1. Cayley-Hamilton Theorem.[1] *Every square matrix A of order n satisfies its characteristic equation; that is, for the characteristic equation*

$$\lambda^n + b_{n-1}\lambda^{n-1} + \cdots + b_1\lambda + b_0 = 0,$$

we have

$$A^n + b_{n-1}A^{n-1} + \cdots + b_1A + b_0I = 0.$$

Example 3. The characteristic equation for Example 1 was found to be $\lambda^2 - 5\lambda + 6 = 0$. According to the Cayley–Hamilton theorem $A^2 - 5A + 6I$ should be 0. To check this we have

$$\begin{bmatrix} 1 & 2 \\ -1 & 4 \end{bmatrix}^2 - 5\begin{bmatrix} 1 & 2 \\ -1 & 4 \end{bmatrix} + 6I$$

$$= \begin{bmatrix} -1 & 10 \\ -5 & 14 \end{bmatrix} - \begin{bmatrix} 5 & 10 \\ -5 & 20 \end{bmatrix} + \begin{bmatrix} 6 & 0 \\ 0 & 6 \end{bmatrix} = \begin{bmatrix} 0 & 0 \\ 0 & 0 \end{bmatrix}.$$

In the next section we are primarily concerned with real symmetric matrices, hence we now state some very important properties concerning real symmetric matrices.

---

[1] A proof of this theorem may be found in *The Algebra of Vectors and Matrices*, T. Wade, Reading, Massachusetts (Addison-Wesley Publishing Company, Inc., 1951), pp. 109–111.

Theorem 8.2. *The characteristic values of real symmetric matrices are real numbers.*

Proof. See Theorem A.10, page 160.

Definition 8.1. *Two real vectors α and β are **orthogonal** if $\alpha \cdot \beta = 0$. A set of real vectors $\{\alpha_1, \alpha_2, \ldots, \alpha_n\}$ is an **orthogonal set** if $\alpha_i \cdot \alpha_j = 0$, where $i \neq j$. If an orthogonal set of vectors has been normalized the set is called an **orthonormal set**.*

Example 4. The symmetric matrix

$$A = \begin{bmatrix} 2 & \sqrt{6} \\ \sqrt{6} & 1 \end{bmatrix}$$

has real characteristic values $\lambda_1 = -1$, $\lambda_2 = 4$, and corresponding characteristic vectors $X_1 = \begin{bmatrix} \sqrt{6} \\ -3 \end{bmatrix}$, $X_2 = \begin{bmatrix} \sqrt{6} \\ 2 \end{bmatrix}$. It can be verified easily that X_1 and X_2 are orthogonal (that is, their dot product is zero). Each of these characteristic vectors can be normalized by dividing each component by the magnitude of the vector. Thus an *orthonormal* set of characteristic vectors would be

$$\begin{bmatrix} \sqrt{6}/\sqrt{15} \\ -3/\sqrt{15} \end{bmatrix} \quad \text{and} \quad \begin{bmatrix} \sqrt{6}/\sqrt{10} \\ 2/\sqrt{10} \end{bmatrix}.$$

The next theorem verifies that it was no accident that the characteristic vectors X_1 and X_2, were orthogonal in the preceding example.

Theorem 8.3. *Any two characteristic vectors X_i, X_j corresponding to two distinct characteristic values λ_i, λ_j of a real symmetric matrix are orthogonal.*

Proof. See Theorem A.11, page 161.

Definition 8.2. *A nonsingular matrix B is an **orthogonal matrix** if $B^{-1} = B^{\mathrm{T}}$.*

Example 5. If

$$B = \begin{bmatrix} \frac{3}{5} & -\frac{4}{5} \\ \frac{4}{5} & \frac{3}{5} \end{bmatrix}, \quad B^{-1} = \frac{\mathrm{adj}\, B}{|B|} = \begin{bmatrix} \frac{3}{5} & \frac{4}{5} \\ -\frac{4}{5} & \frac{3}{5} \end{bmatrix}. \quad \text{Also, } B^{\mathrm{T}} = \begin{bmatrix} \frac{3}{5} & \frac{4}{5} \\ -\frac{4}{5} & \frac{3}{5} \end{bmatrix}.$$

Therefore, B is an orthogonal matrix.

Theorem 8.4.   *A matrix* $T = [X_1 \mid X_2 \mid \ldots \mid X_n]$, *where*

$$\{X_1, X_2, \ldots, X_n\}$$

is an orthonormal set of n-dimensional, real vectors, is an orthogonal matrix. That is, $T^{\mathbf{T}} = T^{-1}$.

Proof.   See Theorem A.12, page 161.

Example 6.   In Example 4, for

$$T = [X_1 \mid X_2] = \begin{bmatrix} \sqrt{6}/\sqrt{15} & \vdots & \sqrt{6}/\sqrt{10} \\ -3/\sqrt{15} & \vdots & 2/\sqrt{10} \end{bmatrix},$$

Theorem 8.4 assures us that T is orthogonal, that is, $T^{-1} = T^{\mathbf{T}}$.

The last two theorems allow us to conclude that *if a matrix T is formed in such a way that the columns are normalized characteristic vectors corresponding to n distinct characteristic values of an n by n real symmetric matrix A, then* $T^{-1} = T^{\mathbf{T}}$. This result proves useful in the next section.

Frequently, characteristic values are called **eigenvalues** or **characteristic roots** or **latent roots**, and characteristic vectors are called **eigenvectors.** (*Eigen* is the German word for characteristic.)

APPLICATIONS

Example 7.   Suppose we perform a certain linear transformation $T = \begin{bmatrix} 1 & 0 \\ 2 & -1 \end{bmatrix}$ on $X = \begin{bmatrix} x_1 \\ x_2 \end{bmatrix}$, that is, $Y = TX$. The question then arises: For which vectors X_i is each image vector Y_i a scalar multiple of X_i? In other words find $X = X_i$ such that $Y = \lambda X$. But since $Y = TX$ the problem becomes find nonzero solutions of $TX = \lambda X$, which is a characteristic value problem. Solutions are characteristic vectors of T, two of which are

$$X_1 = \begin{bmatrix} 1 \\ 1 \end{bmatrix} \text{ and } X_2 = \begin{bmatrix} 0 \\ 1 \end{bmatrix}.$$

Example 8.   Consider the compression of an object of shape *abcd* in Figure 8.1. After compression, the shape is $a'b'c'd'$. Vector Oc' becomes a scalar multiple of Oc and vector Oe' becomes a scalar multiple of Oe; thus Oc and Oe are characteristic vectors; the ratios $\dfrac{|Oc'|}{|Oc|}$ and $\dfrac{|Oe'|}{|Oe|}$ are the corresponding characteristic values under the compression transformation.

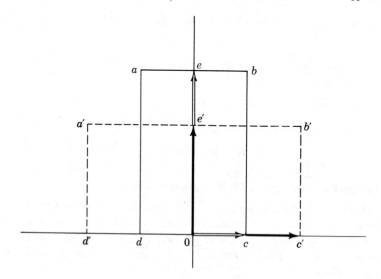

Fig. 8.1

Next consider a compression and a shear as shown in Figure 8.2, where the vector Oc' is a scalar multiple of Oc. Thus, Oc remains a characteristic vector, but notice that Oe no longer is. Figure 8.2 illustrates that under this transformation there exists a point g with an image g' such that Og' is a scalar multiple of Og and hence Og is the other characteristic vector; the two characteristic vectors are no longer orthogonal (perpendicular in two dimensions).

Now in the general two-dimensional case let u_1 and u_2 be normalized, non-collinear, characteristic vectors with corresponding characteristic values λ_1 and λ_2. Prior to the transformation any vector α can be expressed as

$$\alpha = k_1 u_1 + k_2 u_2,$$

and if we assume that the effect of deformation due to the transformation is uniform, then the image is

$$\alpha' = \lambda_1 k_1 u_1 + \lambda_2 k_2 u_2,$$

and hence can easily be located. This is a rather useful idea and is restated for emphasis: An arbitrary vector is resolved into components along non-collinear characteristic vectors; thus the image under the transformation may be considered as a linear combination of those components, and moreover, the scalar multiples are the characteristic values.

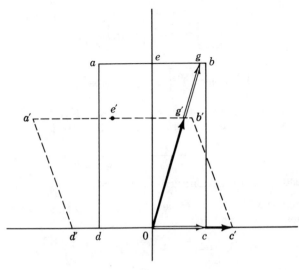

Fig. 8.2

Example 9. Consider some organism or other object which reproduces itself every month, and suppose that the new offspring must wait 2 months before it begins to reproduce. If none die, this situation can be represented by the *difference equation*

$$x_n = x_{n-1} + x_{n-2} \qquad (n = 2, 3, \ldots),$$

where x_n represents the number of objects after n months. Assume that $x_0 = 1$, $x_1 = 1$; then $x_2 = 2$, $x_3 = 3$, $x_4 = 5$, etc. We wish to find a general expression for x_n. Such equations in which the independent variable n must assume certain integral values are called difference equations; they appear in many types of problems dealing with electrical theory, investment problems, the social sciences, statistics, and mechanics. A solution of the given difference equation can be found by use of matrices as follows:

Let $y_{n-1} = x_{n-2}$ or $y_n = x_{n-1}$ be introduced so that we obtain the system of equations

$$\begin{cases} x_n = x_{n-1} + y_{n-1}, \\ y_n = x_{n-1}, \end{cases} \quad \text{or} \quad \begin{bmatrix} x_n \\ y_n \end{bmatrix} = \begin{bmatrix} 1 & 1 \\ 1 & 0 \end{bmatrix} \begin{bmatrix} x_{n-1} \\ y_{n-1} \end{bmatrix}.$$

If $n = 2$,

$$\begin{bmatrix} x_2 \\ y_2 \end{bmatrix} = A \begin{bmatrix} x_1 \\ y_1 \end{bmatrix}, \text{ where } A = \begin{bmatrix} 1 & 1 \\ 1 & 0 \end{bmatrix}.$$

If $n = 3$,

$$\begin{bmatrix} x_3 \\ y_3 \end{bmatrix} = A \begin{bmatrix} x_2 \\ y_2 \end{bmatrix} = A^2 \begin{bmatrix} x_1 \\ y_1 \end{bmatrix}.$$

In general,

$$\begin{bmatrix} x_n \\ y_n \end{bmatrix} = A^{n-1} \begin{bmatrix} x_1 \\ y_1 \end{bmatrix} \qquad \text{for} \qquad n = 2, 3, \ldots.$$

Since we know $x_1 = 1$ and $y_1 = x_0 = 1$, all we need is A^{n-1}.
A formula[2] for finding A^k, where A is a 2 by 2 matrix, is

$$A^k = \frac{\lambda_2 \lambda_1{}^k - \lambda_1 \lambda_2{}^k}{\lambda_2 - \lambda_1} I + \frac{\lambda_2{}^k - \lambda_1{}^k}{\lambda_2 - \lambda_1} A,$$

where λ_1 and λ_2 are distinct characteristic values of A.

In this example $\lambda_1 = \frac{1}{2}(1 + \sqrt{5})$ and $\lambda_2 = \frac{1}{2}(1 - \sqrt{5})$, and after considerable manipulation we find that

$$x_n = \frac{1}{\sqrt{5}} \left[\left(\frac{1 + \sqrt{5}}{2} \right)^{n+1} - \left(\frac{1 - \sqrt{5}}{2} \right)^{n+1} \right].$$

Example 10. This example presupposes some knowledge of calculus and physics. Consider the mechanical system shown in Figure 8.3 where the downward direction is considered to be the positive direction.

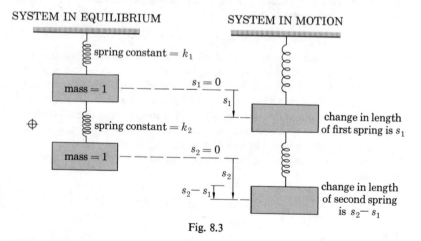

Fig. 8.3

---

[2] A derivation of this formula (which makes use of the Cayley–Hamilton theorem) can be found in P. J. Davis, *The Mathematics of Matrices* (New York, Blaisdell, 1965), pp. 274–275.

The displacements s_1 and s_2 shown in the second diagram are due to a force F, which has been exerted on the system. The well-known physics formulas $F = ms''$ and $F = ks$ allow us to write $ms'' = ks$, where s'' is the acceleration equal to the second derivative of s with respect to t, and m is the mass of the object. Because the mass of each object is one unit in this example, if the damping and the mass of the springs are neglected, the vertical motion of the system is governed by the equations

$$\begin{cases} s_1'' = -k_1 s_1 + k_2(s_2 - s_1), \\ s_2'' = -k_2(s_2 - s_1), \end{cases}$$

or

$$\begin{bmatrix} s_1'' \\ s_2'' \end{bmatrix} = \begin{bmatrix} -(k_1 + k_2) & k_2 \\ k_2 & -k_2 \end{bmatrix} \begin{bmatrix} s_1 \\ s_2 \end{bmatrix}.$$

The first equation is the sum of the force $-k_1 s_1$ exerted on the first body by the first spring (negative if $s_1 > 0$) and the force $k_2(s_2 - s_1)$ exerted on the first body by the second spring (positive if $s_2 - s_1 > 0$). The second equation is the force $-k_2(s_2 - s_1)$ exerted on the second body (negative if $s_2 - s_1 > 0$). In order to solve this system we assume the solution which is obtained by calculus,

$$\begin{bmatrix} s_1 \\ s_2 \end{bmatrix} = \begin{bmatrix} c_1 \\ c_2 \end{bmatrix} e^{\omega t},$$

where c_1 and c_2 are arbitrary constants which are to be determined. Upon differentiating twice we obtain

$$\begin{bmatrix} s_1'' \\ s_2'' \end{bmatrix} = \omega^2 \begin{bmatrix} c_1 \\ c_2 \end{bmatrix} e^{\omega t}.$$

Upon substitution in both sides of the original matrix equation we obtain

$$\omega^2 \begin{bmatrix} c_1 \\ c_2 \end{bmatrix} e^{\omega t} = \begin{bmatrix} -(k_1 + k_2) & k_2 \\ k_2 & -k_2 \end{bmatrix} \begin{bmatrix} c_1 \\ c_2 \end{bmatrix} e^{\omega t},$$

or

$$\omega^2 \begin{bmatrix} c_1 \\ c_2 \end{bmatrix} = \begin{bmatrix} -(k_1 + k_2) & k_2 \\ k_2 & -k_2 \end{bmatrix} \begin{bmatrix} c_1 \\ c_2 \end{bmatrix},$$

which is a characteristic value problem. A solution then is a characteristic vector times $e^{\omega t}$, where ω^2 is the corresponding characteristic value. Notice that the matrix is symmetric; hence ω^2 is real.

EXERCISES

1. Find the characteristic equation, characteristic values, and normalized characteristic vectors for each of the following matrices.

$$(a) \begin{bmatrix} 2 & -2 \\ -2 & 5 \end{bmatrix}; \qquad (b) \begin{bmatrix} 1 & 3 \\ -1 & 5 \end{bmatrix}; \qquad (c) \begin{bmatrix} 5 & 3 \\ -1 & 1 \end{bmatrix};$$

$$(d) \begin{bmatrix} 2 & 1 & 1 \\ 1 & 1 & 0 \\ 1 & 0 & 1 \end{bmatrix}; \qquad (e) \begin{bmatrix} 4 & 3 \\ 1 & 2 \end{bmatrix}; \qquad (f) \begin{bmatrix} 0 & 2 \\ 2 & 0 \end{bmatrix}.$$

2. Find a nonzero solution of the system $AX = \lambda X$ where

$$(a) \ A = \begin{bmatrix} 5 & 1 \\ 4 & 8 \end{bmatrix}; \qquad (b) \ A = \begin{bmatrix} 0 & 3 \\ 1 & 2 \end{bmatrix}; \qquad (c) \ A = \begin{bmatrix} 0 & 0 & 2 \\ 0 & 1 & 0 \\ 2 & 0 & 0 \end{bmatrix}.$$

3. Find three different second-order matrices which have the property that their characteristic values are $\lambda_1 = 1$ and $\lambda_2 = 2$.

4. Find the characteristic values of

$$(a) \begin{bmatrix} 3 & 0 \\ 0 & -4 \end{bmatrix}; \qquad\qquad (b) \begin{bmatrix} a & 0 \\ 0 & b \end{bmatrix};$$

(c) Generalize the result found in (b) to nth-order diagonal matrices.

5. What can be said about the number of characteristic values of an n by n matrix?

6. Verify the Cayley–Hamilton Theorem 8.1 for Exercise 1(a).

7. (a) Find a real matrix whose characteristic values are not real. (b) Find a symmetric matrix whose eigenvalues are not real.

8. Calculate the discriminant of the quadratic characteristic equation of the real matrix $\begin{bmatrix} a & b \\ c & d \end{bmatrix}$. Show that the characteristic values are real and unequal if b and c have the same sign (i.e., $bc > 0$). Under what condition are the characteristic values equal?

9. Apply Theorem 8.3 to each part of Exercise 1 if possible. If not possible, state why it is not.

10. In Exercise 1(a) find an orthonormal set of characteristic vectors. Create a matrix T with the properties stated in Theorem 8.4 and verify that $T^T = T^{-1}$.

11. Find a counterexample to disprove the conjecture that any two characteristic vectors corresponding to distinct characteristic values of a real matrix are orthogonal. How does this conjecture differ from Theorem 8.3?

12. Illustrate Theorem 8.4 with Exercise 1(d).

13. Determine by inspection whether or not each of the following is an orthogonal matrix.

(a) $\begin{bmatrix} \sqrt{2}/2 & -\sqrt{2}/2 \\ \sqrt{2}/2 & \sqrt{2}/2 \end{bmatrix}$;   (b) $\begin{bmatrix} 1 & 0 \\ 0 & 2 \end{bmatrix}$;

(c) $\begin{bmatrix} 1 & 0 & 0 \\ 0 & 1 & 0 \\ 0 & 0 & 1 \end{bmatrix}$;   (d) $\begin{bmatrix} 1 & 0 & 0 \\ 0 & \sqrt{2}/2 & -\sqrt{2}/2 \\ 0 & \sqrt{2}/2 & \sqrt{2}/2 \end{bmatrix}$.

14. Prove or disprove the following conjecture: If a matrix T is formed in such a way that the columns are normalized characteristic vectors corresponding to n distinct characteristic values of an n by n real, symmetric matrix, then T is an orthogonal matrix.

15. For the transformation $\begin{bmatrix} y_1 \\ y_2 \end{bmatrix} = \begin{bmatrix} 1 & 2 \\ -1 & 4 \end{bmatrix} \begin{bmatrix} x_1 \\ x_2 \end{bmatrix}$ find an X such that $Y = \lambda X$.

16. In Figure 8.1 of Example 8 verify that $T = \begin{bmatrix} 2 & 0 \\ 0 & \frac{2}{3} \end{bmatrix}$ is the compression transformation if $a = (-2, 6)$, $b = (2, 6)$, $a' = (-4, 4)$, and $b' = (4, 4)$ and find the characteristic values.

17. In Figure 8.2 of Example 8 verify that $T = \begin{bmatrix} \frac{3}{2} & -\frac{1}{4} \\ 0 & \frac{2}{3} \end{bmatrix}$ is the compression and shear transformation if $a = (-2, 6)$, $b = (2, 6)$, $a' = (-\frac{9}{2}, 4)$, and $b' = (\frac{3}{2}, 4)$. Find the characteristic vectors Oc and Og.

18. As shown in Example 9, solve the difference equation

$$x_n = 6x_{n-1} - 8x_{n-2},$$

where $x_0 = 1$, $x_1 = 2$.

8.2 QUADRATIC FORMS

A quadratic expression which includes only second-degree terms, for example,

$$g(x, y, z) = ax^2 + (2b)xy + cy^2 + (2d)yz + ez^2 + (2f)xz,$$

is called a *quadratic form.* This particular quadratic form can be represented in matrix notation as

$$g(x, y, z) = X^T AX,$$

where

$$X = \begin{bmatrix} x \\ y \\ z \end{bmatrix} \quad \text{and} \quad A = \begin{bmatrix} a & b & f \\ b & c & d \\ f & d & e \end{bmatrix}.$$

Notice that A is symmetric. Geometrically, the equation $g(x, y, z) = k$ can be visualized as a quadric surface with center at the origin. In applied work it frequently is desirable to rotate the axes so that the equation of the surface with respect to the new axes can be expressed as

$$h(u, v, w) = d_1 u^2 + d_2 v^2 + d_3 w^2 = k.$$

In other words, the product (nonsquared) terms are eliminated; that is, the matrix A is diagonalized. The object of this section is to relate this problem to the material of the previous section.

Example 1.   Consider the equation of a given surface:

$$g(x, y, z) = 5x^2 + 3y^2 + 3z^2 - 2xy + 2yz - 2xz = 1$$

or

$$[x \quad y \quad z] \begin{bmatrix} 5 & -1 & -1 \\ -1 & 3 & 1 \\ -1 & 1 & 3 \end{bmatrix} \begin{bmatrix} x \\ y \\ z \end{bmatrix} = 1,$$

or

$$g(x, y, z) = X^T AX = 1.[3]$$

---

[3] The reader may wonder why we say $g(x, y, z) = X^T AX = 1$ rather than $[g(x, y, z)] = X^T AX = [1]$; this is because of the one-to-one correspondence between the set of all scalars and the set of all 1 by 1 matrices with scalar entries, and because this one-to-one correspondence is preserved under addition and multiplication. Further discussion of this may be found in F. E. Hohn, *Elementary Matrix Algebra* (New York, The Macmillan Company, 1958), p. 21.

We wish to perform a linear transformation so that the equation of the surface is of the form

$$h(u, v, w) = d_1u^2 + d_2v^2 + d_3w^2 = 1,$$

or

$$h(u, v, w) = \begin{bmatrix} u & v & w \end{bmatrix} \begin{bmatrix} d_1 & 0 & 0 \\ 0 & d_2 & 0 \\ 0 & 0 & d_3 \end{bmatrix} \begin{bmatrix} u \\ v \\ w \end{bmatrix} = 1,$$

or

$$h(u, v, w) = U^T DU = 1.$$

Suppose we let T represent the transformation matrix with which we hope to eliminate the product terms; that is, let $X = TU$. Then under this transformation,

$$X^T AX = (TU)^T A(TU) = U^T(T^T AT)U.$$

Thus we are looking for a matrix T, if it exists, such that $T^T AT = D$ is a diagonal matrix, and, of course, we would like to know what this diagonal matrix is.

Theorem 8.5. *For a real symmetric n by n matrix A with distinct characteristic values λ_i and corresponding normalized characteristic vectors X_i, there exists a transformation matrix*

$$T = [X_1 \mid X_2 \mid \ldots \mid X_n]$$

such that

$$T^T AT = \begin{bmatrix} \lambda_1 & 0 & \cdots & 0 \\ 0 & \lambda_2 & \cdots & 0 \\ \cdot & \cdot & \cdot & \cdot \\ \cdot & \cdot & \cdot & \cdot \\ \cdot & \cdot & \cdot & \cdot \\ 0 & 0 & \cdots & \lambda_n \end{bmatrix}.$$

Proof. By matrix multiplication and because it is given that X_i are characteristic vectors of A corresponding to λ_i, we have

$$AT = A[X_1 \mid X_2 \mid \ldots \mid X_n] = [AX_1 \mid AX_2 \mid \ldots \mid AX_n]$$
$$= [\lambda_1 X_1 \mid \lambda_2 X_2 \mid \ldots \mid \lambda_n X_n].$$

By Exercise 7 of Section 2.3 and matrix multiplication the last expression is equal to

$$[X_1\lambda_1 \mid X_2\lambda_2 \mid \cdots \mid X_n\lambda_n] = [X_1 \mid X_2 \mid \cdots \mid X_n] \begin{bmatrix} \lambda_1 & \cdots & 0 \\ & \lambda_2 & & \\ & & \ddots & \\ 0 & \cdots & & \lambda_n \end{bmatrix}$$

$$= T \begin{bmatrix} \lambda_1 & \cdots & 0 \\ & \lambda_2 & & \\ & & \ddots & \\ 0 & \cdots & & \lambda_n \end{bmatrix} .$$

Premultiplying by T^{-1}, which we know exists by Theorem 8.4, yields

$$T^{-1}AT = T^{-1}T \begin{bmatrix} \lambda_1 & \cdots & 0 \\ & \lambda_2 & & \\ & & \ddots & \\ 0 & \cdots & & \lambda_n \end{bmatrix} = \begin{bmatrix} \lambda_1 & \cdots & 0 \\ & \lambda_2 & & \\ & & \ddots & \\ 0 & \cdots & & \lambda_n \end{bmatrix} .$$

Hence,

$$T^{\mathbf{T}}AT = \begin{bmatrix} \lambda_1 & \cdots & 0 \\ & \lambda_2 & & \\ & & \ddots & \\ 0 & \cdots & & \lambda_n \end{bmatrix}$$

by Theorems 8.3 and 8.4.

Example 2. Referring back to Example 1 we make use of Theorem 8.5 to say that the diagonal matrix D has the characteristic values of A on the main diagonal. These characteristic values can be found to be 2, 3, and 6.

Thus,

$$5x^2 + 3y^2 + 3z^2 - 2xy + 2yz - 2xz = 1$$

becomes

$$2u^2 + 3v^2 + 6w^2 = 1$$

under the rotation transformation

$$T = \begin{bmatrix} 1/\sqrt{2} & 1/\sqrt{3} & 1/\sqrt{6} \\ 0 & 1/\sqrt{3} & -2/\sqrt{6} \\ -1/\sqrt{2} & 1/\sqrt{3} & 1/\sqrt{6} \end{bmatrix}.$$

The reader should note that neither T nor D is unique because the eigenvalues may be listed in any order (however, for a rigid rotation det $T = +1$).

In connection with the preceding material of this section, the following terminology is customary, and is therefore introduced for the reader's benefit.

Definition 8.3. *Matrix A is said to be **congruent** to matrix B if there exists a nonsingular matrix P such that $B = P^T A P$.*

Example 3. $\begin{bmatrix} 1 & 4 \\ 2 & 5 \end{bmatrix}$ is congruent to $\begin{bmatrix} 1 & 2 \\ 0 & -3 \end{bmatrix}$ because there exists a non-singular matrix $P = \begin{bmatrix} 1 & 2 \\ 0 & 1 \end{bmatrix}$ such that

$$\begin{bmatrix} 1 & 2 \\ 0 & -3 \end{bmatrix} = P^T \begin{bmatrix} 1 & 4 \\ 2 & 5 \end{bmatrix} P.$$

Definition 8.4. *Matrix A is said to be **similar** to matrix B if there exists a nonsingular matrix P such that $B = P^{-1}AP$.*

Example 4. $\begin{bmatrix} 1 & 2 \\ -1 & 4 \end{bmatrix}$ is similar to $\begin{bmatrix} 2 & 0 \\ 0 & 3 \end{bmatrix}$ because there exists a non-singular matrix $P = \begin{bmatrix} 2 & -3 \\ 1 & -3 \end{bmatrix}$ such that

$$\begin{bmatrix} 2 & 0 \\ 0 & 3 \end{bmatrix} = P^{-1} \begin{bmatrix} 1 & 2 \\ -1 & 4 \end{bmatrix} P.$$

Definition 8.5. *Matrix A is said to be **orthogonally similar** (or orthogonally congruent) to matrix B if there exists an orthogonal matrix P such that $B = P^{-1}AP = P^T A P$.*

Example 5. The matrix $\begin{bmatrix} 1 & 2 \\ 2 & 1 \end{bmatrix}$ is orthogonally similar to $\begin{bmatrix} -1 & 0 \\ 0 & 3 \end{bmatrix}$ be-

cause there exists a matrix $P = \begin{bmatrix} 1/\sqrt{2} & 1/\sqrt{2} \\ -1/\sqrt{2} & 1/\sqrt{2} \end{bmatrix}$ such that

$$\begin{bmatrix} -1 & 0 \\ 0 & 3 \end{bmatrix} = P^{-1} \begin{bmatrix} 1 & 2 \\ 2 & 1 \end{bmatrix} P = P^{\mathrm{T}} \begin{bmatrix} 1 & 2 \\ 2 & 1 \end{bmatrix} P.$$

Definition 8.6. *A matrix* $B = [X_1 \mid X_2 \mid \cdots \mid X_n]$, *where* X_i *are characteristic vectors corresponding to the n characteristic values of A, is said to be a modal matrix of A.* (Note that B is not unique).

Example 6. Characteristic vectors of $A = \begin{bmatrix} 1 & 2 \\ -1 & 4 \end{bmatrix}$ are $\begin{bmatrix} 2 \\ 1 \end{bmatrix}$ and $\begin{bmatrix} -3 \\ -3 \end{bmatrix}$ corresponding to the characteristic values 2 and 3; therefore a modal matrix of A is $\begin{bmatrix} 2 & -3 \\ 1 & -3 \end{bmatrix}$.

Notice that in Examples 1 and 2 of this section, the transformation matrix T is an orthogonal, modal matrix of A, as required by Theorem 8.5. Moreover A and D are similar, congruent, and orthogonally similar.

It can be shown that two matrices which are similar have determinants which are the same (Exercise 8); also their traces and characteristic roots are the same (Exercises 9 and 10); we say that these quantities are *invariant* under the similarity transformation.

The reader should note that the hypothesis of Theorem 8.5 required that the matrix A of the quadratic form $X^{\mathrm{T}}AX$ be symmetric. Why? (Exercise 6). It should also be observed that we required that the characteristic values of the symmetric matrix be distinct. The latter restriction can be avoided,[4] and although we do not discuss the theory, the following example illustrates the procedure.

Example 7. For the quadratic form

$$f(x, y, z) = X^{\mathrm{T}}AX = X^{\mathrm{T}} \begin{bmatrix} 2 & 0 & 1 \\ 0 & 3 & 0 \\ 1 & 0 & 2 \end{bmatrix} X,$$

---

[4] For a justification see G. Hadley, *Linear Algebra* (Reading, Massachusetts, Addison-Wesley Publishing Company, Inc., 1961), pp. 243–248.

the characteristic values of A are $\lambda_1 = 1$, $\lambda_2 = \lambda_3 = 3$. A normalized charac-

teristic vector corresponding to $\lambda_1 = 1$ is found to be $X_1 = \begin{bmatrix} 1/\sqrt{2} \\ 0 \\ -1/\sqrt{2} \end{bmatrix}$. The

characteristic values $\lambda_2 = \lambda_3 = 3$ yield

$$(A - 3I)X = 0 \qquad \text{or} \qquad \begin{bmatrix} -1 & 0 & 1 \\ 0 & 0 & 0 \\ 1 & 0 & -1 \end{bmatrix} \begin{bmatrix} x \\ y \\ z \end{bmatrix} = \begin{bmatrix} 0 \\ 0 \\ 0 \end{bmatrix},$$

whose complete solution is

$$\begin{cases} x = z, \\ y \quad \text{arbitrary}; \end{cases}$$

therefore, we must choose two vectors subject to these conditions and such that they are orthogonal to X_1 and to each other. The vectors $\begin{bmatrix} 1 \\ 1 \\ 1 \end{bmatrix}$ and $\begin{bmatrix} 1 \\ -2 \\ 1 \end{bmatrix}$ suffice; notice that they are not unique. Normalized, these vectors become

$$X_2 = \begin{bmatrix} 1/\sqrt{3} \\ 1/\sqrt{3} \\ 1/\sqrt{3} \end{bmatrix} \text{ and } X_3 = \begin{bmatrix} 1/\sqrt{6} \\ -2/\sqrt{6} \\ 1/\sqrt{6} \end{bmatrix}. \text{ Therefore,}$$

$$T = \begin{bmatrix} 1/\sqrt{2} & 1/\sqrt{3} & 1/\sqrt{6} \\ 0 & 1/\sqrt{3} & -2/\sqrt{6} \\ -1/\sqrt{2} & 1/\sqrt{3} & 1/\sqrt{6} \end{bmatrix} \qquad \text{and} \qquad T^\mathsf{T}AT = \begin{bmatrix} 1 & 0 & 0 \\ 0 & 3 & 0 \\ 0 & 0 & 3 \end{bmatrix}.$$

In this chapter we have been concerned primarily with characteristic value problems involving real, symmetric matrices. Although we do not wish to investigate in depth the corresponding problems involving real, nonsymmetric matrices, it may be enlightening to point out a few of the differences:

(1) Characteristic values of a real, nonsymmetric matrix A need not be real.

(2) Characteristic vectors of a real, nonsymmetric matrix A need not be orthogonal. (See Example 6 of this section for an illustration.)

(3) A similarity transformation matrix T, such that $T^{-1}AT$ is a diagonal matrix, need not exist for a real, nonsymmetric matrix A. If there exists a nonsingular modal matrix P of A then $P^{-1}AP$ is a diagonal matrix; such a P must exist if the characteristic values

of A are distinct. When T does exist such that $T^{-1}AT$ is a diagonal matrix, the main diagonal of $T^{-1}AT$ consists of the characteristic values.

(4) When a similarity transformation matrix T does exist such that $T^{-1}AT$ is a diagonal matrix for a real, nonsymmetric A then T need not be orthogonal; that is, T^{-1} need not equal T^T. (Examples 4 and 6 of this section illustrate this point.)

APPLICATIONS

Applications making use of the material covered in this section are legion but many of these applications are quite sophisticated. The textbook *Matrix Methods for Engineering*, by L. A. Pipes (Englewood Cliffs, New Jersey, Prentice-Hall, Inc., 1963), offers scores of illustrations of the previous statement. Applications of this material are not limited to engineering, however, as we will demonstrate in Examples 8, 10, and 11. A few specific examples of the many uses of the concepts of this section are now given.

Example 8. Consider the symmetric correlation matrix of Example 9 in Section 1.2; such matrices are of considerable importance in the branch of statistics known as factor analysis. One of the primary problems of factor analysis involves the diagonalization of matrices of correlation.[5] Although psychologists did much of the original work in this area, there is considerable potential for its usefulness in other areas of the social sciences such as sociology and political science.[6]

Example 9. Consider a rigid body in motion with respect to an X-coordinate system, and let a moving coordinate system Y be fixed in the body with the origin at the center of mass. We may regard the rigid body as a set of particles with mass m, whose relative distances from each other are constant. The symmetric matrix

$$J = \begin{bmatrix} m(y_2{}^2 + y_3{}^2) & -my_1y_2 & -my_1y_3 \\ -my_1y_2 & m(y_1{}^2 + y_3{}^2) & -my_2y_3 \\ -my_1y_3 & -my_2y_3 & m(y_1{}^2 + y_2{}^2) \end{bmatrix}$$

[5] L. L. Thurstone, *Multiple Factor Analysis* (Chicago, University of Chicago Press, 1947), Chap. 20. A later reference is D. N. Lawley and A. E. Maxwell, *Factor Analysis as a Statistical Method* (London, Butterworths, 1963).

[6] For instance, see H. Alker, Jr., "Dimensions of Conflict in the General Assembly," *American Political Science Review*, Vol. 58 (1964), pp. 642–657.

is called the *inertia matrix* of the rigid body. The entries on the main diagonal represent the moments of inertia with respect to the coordinate axes y_1, y_2, and y_3. The other entries represent the products of inertia of the body. It is desirable to introduce a new set of axes z_1, z_2, and z_3 by a rotation transformation such that the products of inertia vanish. The transformation matrix R (such that $Y = RZ$) has as its columns the normalized characteristic vectors of J. The resulting inertia matrix with respect to the Z-coordinate system is

$$J_z = \begin{bmatrix} m(z_2^2 + z_3^2) & 0 & 0 \\ 0 & m(z_1^2 + z_3^2) & 0 \\ 0 & 0 & m(z_1^2 + z_2^2) \end{bmatrix}.$$

The entries on the main diagonal are the characteristic values of J and are called the principal moments of inertia. The axes z_1, z_2, and z_3 are called the principal axes of inertia.

Example 10. Some of the concepts mentioned in this chapter are useful in the study of various stability problems in economics. One problem is to consider the stability of a system of difference equations, such as

$$\begin{cases} x_n = a_{11}x_{n-1} + a_{12}y_{n-1}, \\ y_n = a_{21}x_{n-1} + a_{22}y_{n-1}. \end{cases}$$

Perhaps x_n and y_n represent changes in certain economic indices after n intervals of time, and the question is whether or not these changes are finite as n gets arbitrarily large. We rewrite the system using matrix notation,

$$\begin{bmatrix} x_n \\ y_n \end{bmatrix} = A \begin{bmatrix} x_{n-1} \\ y_{n-1} \end{bmatrix}, \qquad \text{or} \qquad X_n = AX_{n-1},$$

and as we found in Example 9 of Section 8.1 the latter statement can be written as

$$X_n = A^{n-1}X_1.$$

Now if a similarity transformation matrix T exists such that $T^{-1}AT = D$ is a diagonal matrix, then

$$A = TDT^{-1},$$

$$A^2 = (TDT^{-1})(TDT^{-1}) = TD^2T^{-1},$$

$$\cdot$$
$$\cdot$$
$$\cdot$$

$$A^{n-1} = TD^{n-1}T^{-1}.$$

Therefore,

$$X_n = A^{n-1}X_1 = (TD^{n-1}T^{-1})X_1$$

$$= T\begin{bmatrix} \lambda_1^{n-1} & 0 \\ 0 & \lambda_2^{n-1} \end{bmatrix} T^{-1}X_1,$$

where the λ_i are the characteristic values of A, and T is a modal matrix of A. The matrix X_1 represents the initial conditions. (The last equation offers an alternative method of solving Example 9 of Section 8.1.) From the last equation we can see that the solutions are of the form

$$\begin{cases} x_n = c_1\lambda_1^{n-1} + c_2\lambda_2^{n-1}, \\ y_n = c_3\lambda_1^{n-1} + c_4\lambda_2^{n-1}, \end{cases}$$

where the c_i are known constants. If $|\lambda_i| < 1$, then x_n and y_n must converge to zero (achieve stability) as $n \to \infty$.

Example 11. Recall that it was pointed out earlier in this section that the determinant of a matrix and the trace of a matrix are invariant under a similarity transformation. We can use this knowledge in the following way: It can be proved that a matrix A, whose characteristic values λ_i are distinct, is similar to a diagonal matrix D, where the diagonal elements of D are the characteristic values. (See Exercise 7(b) of this section.) Therefore for such an n by n matrix A (where $\Rightarrow$ means implies):

(1) $|\lambda_i| < 1$ for all $i \Rightarrow \left| \sum\limits_{i=1}^{n} \lambda_i \right| < n$

$\Rightarrow |\operatorname{tr} D| < n$

$\Rightarrow |\operatorname{tr} A| < n \Rightarrow -n < \operatorname{tr} A < n.$

(2) $\lambda_i < 0$ for all $i \Rightarrow \sum\limits_{i=1}^{n} \lambda_i < 0$

$\Rightarrow \operatorname{tr} D < 0 \Rightarrow \operatorname{tr} A < 0.$

(3) $|\lambda_i| < 1$ for all $i \Rightarrow |\det D| < 1$

$\Rightarrow |\det A| < 1 \Rightarrow -1 < \det A < 1.$

(4) $\lambda_i < 0$ for all $i \Rightarrow \begin{cases} \det D < 0 \text{ if } n \text{ is odd,} \\ \det D > 0 \text{ if } n \text{ is even.} \end{cases}$

$\Rightarrow \begin{cases} \det A < 0 \text{ if } n \text{ is odd,} \\ \det A > 0 \text{ if } n \text{ is even.} \end{cases}$

These implications and the contrapositives of these implications are sometimes useful in stability theory.

The reader can acquire a further appreciation of the usefulness of the material of this chapter by consulting B. Noble, *Applications of*

Undergraduate Mathematics in Engineering (New York, The Macmillan Company, 1967), pp. 44–65, 102–110, and 251–265. The numerous applications found therein arise from realistic engineering problems and are explained in some detail.

EXERCISES

1. In the following quadratic equations eliminate the product terms (that is, the terms involving xy, xz, or yz) by finding and using an orthogonal modal matrix T.

(a) $2x^2 - 4xy + 5y^2 = 6$, which can be expressed as

$$[x \ \ y] \begin{bmatrix} 2 & -2 \\ -2 & 5 \end{bmatrix} \begin{bmatrix} x \\ y \end{bmatrix} = 6;$$

(b) $x^2 + 4xy + y^2 = 1$;

(c) $2x^2 + 12xy - 3y^2 = 1$;

(d) $2x^2 + y^2 + z^2 + 2xy + 2xz = 12$ (*Hint:* Use Exercise 1(*d*) of Section 8.1);

(e) $x^2 + 4y^2 + 3z^2 - 4xy = 15$;

(f) $xy = 4$.

2. A primary problem in analytic geometry is to rotate the axes so that the equation of a given conic is in standard form. Demonstrate that this is what was accomplished in Exercise 1 (*f*) by graphing the given equation and the resulting equation in the answer.

3. If we find a matrix T such that $T^{-1}AT$ is a diagonal matrix, we say that A has been diagonalized. Diagonalize the following matrices, if possible. If it is not possible, state why it is not possible.

(a) $\begin{bmatrix} 1 & 2 \\ 2 & 1 \end{bmatrix}$;   (b) $\begin{bmatrix} 1 & -1 \\ 1 & 3 \end{bmatrix}$;   (c) $\begin{bmatrix} 1 & 0 & 1 \\ 0 & 1 & 2 \\ 1 & 2 & 5 \end{bmatrix}$.

4. Find a transformation matrix T such that T^TAT is equal to a diagonal matrix, and then find that diagonal matrix.

$$A = \begin{bmatrix} 2 & -1 & 1 \\ -1 & 2 & -1 \\ 1 & -1 & 2 \end{bmatrix}.$$

(*Hint:* See Example 7 of this section.)

5. (*a*) What can one say about the stability of the following system:

$$\begin{cases} x_n = x_{n-1} - \tfrac{1}{6}y_{n-1}, \\ y_n = x_{n-1} + \tfrac{1}{6}y_{n-1}? \end{cases}$$

(*b*) What can one say about the characteristic values of the coefficient matrix of this system by considering the trace and determinant of that matrix?

6. Why is A required to be symmetric in Theorem 8.5? (*Hint:* Consider proofs of Theorems 8.3–8.5.)

7. Conjecture: A real matrix A with distinct characteristic values is similar to a diagonal matrix.

(*a*) How does this conjecture differ from Theorem 8.5?

(*b*) Prove or disprove this conjecture given that there exists a non-singular modal matrix P of the matrix A.

8. Prove that if A is similar to B, then $|A| = |B|$. (*Hint:* Remember $|P| \, |P^{-1}| = 1$.)

9. Prove that if A is similar to B, then the characteristic values of the two matrices are equal. (*Hint:* Show that $|B - \lambda I| = |A - \lambda I|$, and let $I = P^{-1}IP$.)

10. Prove that if A is similar to B, then tr A = tr B. (*Hint:* Use Theorem 3.8.)

11. In Example 10 of this section it was found that if a similarity transformation matrix T exists, then

$$X_n = A^{n-1}X_1 = T\begin{bmatrix} \lambda_1^{n-1} & 0 \\ 0 & \lambda_2^{n-1} \end{bmatrix}T^{-1}X_1.$$

Use this result to solve Exercise 18 of Section 8.1.

12. Show that

$$\begin{bmatrix} 3 & 2 \\ 4 & 1 \end{bmatrix} \quad \text{and} \quad \begin{bmatrix} 5 & 1 \\ 3 & 2 \end{bmatrix}$$

are not similar matrices. (*Hint:* See Exercise 8.)

13. A relation R on a set S is said to be an *equivalence relation* if aRa, and $aRb \Rightarrow bRa$, and $(aRb$ and $bRc) \Rightarrow aRc$, where a, b, c belong to S. (*Example:* The relation "equals" is an equivalence relation on the set $\{1, 2, 3, 4\}$ since for any a, b, c in that set $a = a$, and if $a = b$ then $b = a$, and if $a = b$ and $b = c$ then $a = c$.)

(*a*) Prove or disprove that the relation of congruence is an equivalence relation on the set of all *n*th-order, real matrices.

(*b*) Prove or disprove that the relation of similarity is an equivalence relation on the set of all *n*th-order real matrices.

NEW VOCABULARY

§8.1	characteristic equation	§8.1	eigenvalues
§8.1	characteristic values	§8.1	latent roots
§8.1	characteristic vectors	§8.1	characteristic roots
§8.1	normalized vector	§8.1	eigenvectors
§8.1	characteristic function	§8.1	difference equation
§8.1	characteristic polynomial	§8.2	quadratic form
		§8.2	congruent matrices
§8.1	orthogonal vectors	§8.2	similar matrices
§8.1	orthogonal set of vectors	§8.2	orthogonally similar matrices
§8.1	orthonormal set of vectors	§8.2	modal matrix
§8.1	orthogonal matrix	§8.2	invariant

Appendix

A.1 SIGMA NOTATION

For general discussions of matrix multiplication the so-called "Σ notation" (read "sigma notation") is very helpful. Σ is a letter from the Greek alphabet and in mathematics usually stands for the "sum of." For example, the sum

$$1^2 + 2^2 + 3^2 + 4^2 + 5^2 = \sum_{k=1}^{5} k^2.$$

This expression in words is

"the sum of k^2, where k ranges from 1 through 5."

k is called the *index of summation*.

Example 1.

(a) $2 + 4 + 6 + 8 + \cdots + 98 + 100 = \sum_{k=1}^{50} 2k$.

(b) $x_1 + x_2 + x_3 + \cdots + x_{100} = \sum_{k=1}^{100} x_k$.

(c) $a_{i1} + a_{i2} + a_{i3} = \sum_{k=1}^{3} a_{ik}$.

(d) $a_{11}b_{11} + a_{12}b_{21} + a_{13}b_{31} = \sum_{k=1}^{3} a_{1k}b_{k1}$.

(e) $a_{i1}b_{1j} + a_{i2}b_{2j} + a_{i3}b_{3j} = \sum_{k=1}^{3} a_{ik}b_{kj}$.

Example 1(d) may be recognized as the entry in the first row and first column of the matrix product

$$\begin{bmatrix} a_{11} & a_{12} & a_{13} \\ \cdot & \cdot & \cdot \\ \cdot & \cdot & \cdot \\ \cdot & \cdot & \cdot \\ a_{m1} & a_{m2} & a_{m3} \end{bmatrix} \begin{bmatrix} b_{11} & \cdots & b_{1n} \\ b_{21} & \cdots & b_{2n} \\ b_{31} & \cdots & b_{3n} \end{bmatrix} = C,$$

and Example 1(e) is the entry c_{ij}.

151

Example 2. Express the product of two 2 by 2 matrices using $\sum$ notation.

$$\begin{bmatrix} a_{11} & a_{12} \\ a_{21} & a_{22} \end{bmatrix} \begin{bmatrix} b_{11} & b_{12} \\ b_{21} & b_{22} \end{bmatrix} = \begin{bmatrix} (a_{11}b_{11} + a_{12}b_{21}) & (a_{11}b_{12} + a_{12}b_{22}) \\ (a_{21}b_{11} + a_{22}b_{21}) & (a_{21}b_{12} + a_{22}b_{22}) \end{bmatrix}$$

$$= \begin{bmatrix} \sum_{k=1}^{2} a_{1k}b_{k1} & \sum_{k=1}^{2} a_{1k}b_{k2} \\ \sum_{k=1}^{2} a_{2k}b_{k1} & \sum_{k=1}^{2} a_{2k}b_{k2} \end{bmatrix}$$

$$= \begin{bmatrix} \sum_{k=1}^{2} a_{ik}b_{kj} \end{bmatrix}_{(2,2)} .$$

It is helpful to list some of the rules pertaining to the summation notation. First of all, any letter not used for another purpose in the same set of expressions may be used as an index of summation, that is, the following expressions have the same meaning.

$$\sum_{k=1}^{5} k^2 = \sum_{i=1}^{5} i^2.$$

Any factor not involving the index of summation may be moved in front of the $\sum$ sign, that is,

$$\sum_{k=1}^{n} cx_k = c \sum_{k=1}^{n} x_k.$$

Also,

$$\sum_{k=1}^{n} a_k + \sum_{k=1}^{n} b_k = \sum_{k=1}^{n} (a_k + b_k).$$

A.2   DOUBLE SUMMATION

Two summations may occur in succession. The notation

$$\sum_{k=1}^{n} \sum_{i=1}^{m} a_{ik} = \sum_{k=1}^{n} \left(\sum_{i=1}^{m} a_{ik} \right)$$

means that the summation using the index i is to be performed first, giving

$$\sum_{k=1}^{n} (a_{1k} + a_{2k} + a_{3k} + \cdots + a_{mk}).$$

Then the second summation is performed using the index k. We have

$$\sum_{k=1}^{n} \sum_{i=1}^{m} a_{ik} = a_{11} + a_{21} + a_{31} + \cdots + a_{m1}$$
$$+ a_{12} + a_{22} + a_{32} + \cdots + a_{m2}$$
$$\cdot$$
$$\cdot$$
$$\cdot$$
$$+ a_{1n} + a_{2n} + a_{3n} + \cdots + a_{mn}.$$

EXERCISES

Write the following without $\sum$ notation:

1. $\displaystyle\sum_{k=1}^{5} k$.

2. $\displaystyle\sum_{k=3}^{7} (k-2)$.

3. $\displaystyle\sum_{i=1}^{3} i^3$.

4. $\displaystyle\sum_{k=1}^{4} a_k$.

5. $\displaystyle\sum_{k=1}^{3} a_{2k}a_{k3}$.

In Exercises 6–11, express the sums in $\sum$ notation:

6. $3 + 6 + 9 + 12$.

7. $a_{21} + a_{22} + a_{23} + a_{24} + a_{25}$.

8. $a_{21}b_{13} + a_{22}b_{23} + a_{23}b_{33}$.

9. $[a_{11}\ a_{12}\ a_{13}] \begin{bmatrix} b_{11} \\ b_{21} \\ b_{31} \end{bmatrix}$.

10. $\begin{bmatrix} a_{11} & a_{12} \\ a_{21} & a_{22} \end{bmatrix} \begin{bmatrix} b_{11} \\ b_{21} \end{bmatrix}$.

11. $\begin{bmatrix} a_{11} & a_{12} \\ a_{21} & a_{22} \end{bmatrix} \begin{bmatrix} b_{11} & b_{12} & b_{13} \\ b_{21} & b_{22} & b_{23} \end{bmatrix}$.

12. Does $c\displaystyle\sum_{i=1}^{n} (i+4) = \sum_{k=1}^{n} c(k+4)$? Why?

13. Express $\displaystyle\sum_{k=1}^{2} \sum_{i=1}^{2} a_{ik}$ without $\sum$ notation.

14. Express $\displaystyle\sum_{k=1}^{2} \sum_{h=1}^{2} (a_{1h}b_{hk}c_{k1})$ without $\sum$ notation.

A.3   DOT PRODUCT OF TWO VECTORS

Theorem A.1 [Theorem 1.1, page 14]. *If α and β are two nonzero vectors in the x_1x_2-plane, then $\alpha \cdot \beta = |\alpha||\beta| \cos \theta$, where θ is the angle between α and β.* (See Example 5, Section 1.3., page 15).

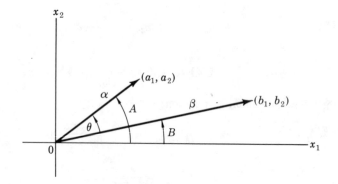

Fig. A.1

Proof.   Refer to Figure A.1.
$$\theta = A - B.$$
$$\cos \theta = \cos (A - B)$$
$$= \cos A \cos B + \sin A \sin B$$
$$= \frac{a_1}{|\alpha|} \frac{b_1}{|\beta|} + \frac{a_2}{|\alpha|} \frac{b_2}{|\beta|}.$$
$$|\alpha||\beta| \cos \theta = a_1 b_1 + a_2 b_2.$$
By Definition 1.10,
$$|\alpha||\beta| \cos \theta = \alpha \cdot \beta.$$

A.4   MATRIX MULTIPLICATION IS ASSOCIATIVE

Theorem A.2 [Theorem 2.4, page 30]. Associative Property. *Given that the three matrices with scalar entries A, B, and C are conformable for multiplication, $(AB)C = A(BC)$.*

Proof. Using the $\sum$ notation, let AB and BC be represented, respectively, by

$$\left[\sum_{h=1}^{p} a_{ih}b_{hk} \right]_{(m,n)} \quad \text{and} \quad \left[\sum_{k=1}^{n} b_{hk}c_{kj} \right]_{(p,q)}.$$

In double summation notation, then, $(AB)C$ and $A(BC)$ become, respectively,

$$\left[\sum_{k=1}^{n} \left(\sum_{h=1}^{p} a_{ih}b_{hk} \right)(c_{kj}) \right]_{(m,q)} \quad \text{and} \quad \left[\sum_{h=1}^{p} (a_{ih}) \left(\sum_{k=1}^{n} b_{hk}c_{kj} \right) \right]_{(m,q)}.$$

If these two expressions are expanded it is seen that they are equal.

A.5 MATRIX MULTIPLICATION IS DISTRIBUTIVE WITH RESPECT TO ADDITION

Theorem A.3. [Theorem 2.5, page 30]. *Assuming conformability and that A, B, C are matrices with scalar entries, then*

$$A(B + C) = AB + AC \qquad \text{(left distributive law)},$$

and

$$(A + B)C = AC + BC \qquad \text{(right distributive law)}.$$

Proof of $A(B + C) = AB + AC$. By definition of addition of matrices and the use of the $\sum$ notation for matrix multiplication let $A(B + C)$ be represented by

$$\left[\sum_{k=1}^{p} a_{ik}(b_{kj} + c_{kj}) \right]_{(m,n)}.$$

By the left distributive law for scalars and by a previously discussed property of the $\sum$ notation, this can be written as

$$\left[\sum_{k=1}^{p} a_{ik}b_{kj} + \sum_{k=1}^{p} a_{ik}c_{kj} \right]_{(m,n)}$$

or

$$\left[\sum_{k=1}^{p} a_{ik}b_{kj} \right]_{(m,n)} + \left[\sum_{k=1}^{p} a_{ik}c_{kj} \right]_{(m,n)},$$

which is the $\sum$ notation for $AB + AC$.

A.6 TRACE OF PRODUCTS

Theorem A.4. [Theorem 3.8, page 55]. *If A and B are n by n matrices, then* tr AB = tr BA.

Proof. Let

$$C = AB = \left[\sum_{k=1}^{n} a_{ik}b_{kj} \right]_{(n,n)}$$

and

$$D = BA = \left[\sum_{k=1}^{n} b_{ik}a_{kj} \right]_{(n,n)}.$$

Then

$$\text{tr } C = \sum_{i=1}^{n} c_{ii} = \sum_{i=1}^{n} \left(\sum_{k=1}^{n} a_{ik}b_{ki} \right),$$

and

$$\text{tr } D = \sum_{i=1}^{n} d_{ii} = \sum_{i=1}^{n} \left(\sum_{k=1}^{n} b_{ik}a_{ki} \right);$$

their expansions are equal.

A.7 CRAMER'S RULE

Theorem A.5. [Theorem 3.16, page 62]. Cramer's Rule. *For the system AX = B, let ^{j}A denote the matrix obtained from A by replacing the jth column of A by the vector B. If* det $A \neq 0$, *then the system AX = B has exactly one solution; this solution is*

$$x_j = \frac{\det (^{j}A)}{\det A}, \qquad j = 1, 2, \ldots, n.$$

(*Note:* An understanding of the material in Section 5.1 is needed for this proof.)

Proof. For the system $AX = B$, where A is nonsingular, we can show that $X = A^{-1}B$ (see page 85). By Theorem 5.1, by definition of the adjoint matrix, and by matrix multiplication

$$X = A^{-1}B = \frac{\text{adj } A}{|A|} B = \frac{1}{|A|} (\text{cof } A)^{\text{T}}B$$

$$
= \frac{1}{|A|}
\begin{bmatrix}
A_{11} & A_{12} & \cdots & A_{1n} \\
A_{21} & A_{22} & \cdots & A_{2n} \\
 & & \cdot & \\
 & & \cdot & \\
 & & \cdot & \\
A_{n1} & A_{n2} & \cdots & A_{nn}
\end{bmatrix}^{\mathrm{T}}
\begin{bmatrix}
b_1 \\ b_2 \\ \cdot \\ \cdot \\ \cdot \\ b_n
\end{bmatrix}
$$

$$
= \frac{1}{|A|}
\begin{bmatrix}
b_1 A_{11} + b_2 A_{21} + \cdots + b_n A_{n1} \\
b_1 A_{12} + b_2 A_{22} + \cdots + b_n A_{n2} \\
\cdot \\ \cdot \\ \cdot \\
b_1 A_{1n} + b_2 A_{2n} + \cdots + b_n A_{nn}
\end{bmatrix}
$$

$$
= \frac{1}{\det A}
\begin{bmatrix}
\det (^1 A) \\
\det (^2 A) \\
\cdot \\ \cdot \\ \cdot \\
\det (^n A)
\end{bmatrix}
\quad \text{by definition of det } (^i A),
$$

or,

$$
x_j = \frac{\det (^i A)}{\det A} \text{ for } j = 1, 2, \cdots, n.
$$

A.8 CONSISTENCY OF A LINEAR SYSTEM

Theorem A.6 [Theorem 4.1, page 73]. *A system of linear equations* $AX = B$ *is consistent if and only if the rank of the augmented matrix is equal to the rank of the coefficient matrix. This common value of the rank of these two matrices, if it exists, is denoted by* r *and is called the rank of the system.*

(*Note:* An understanding of the material in Section 5.2 is needed for this proof.)

Proof. By performing the same suitable elementary row transformations on both $AX = B$ and $[A \mathbin{\vdots} B]$ and, if necessary, properly rearranging the subscripts of the unknowns, we eventually obtain the equivalent system:

$$\begin{cases}
x_1 + c_{1,k+1}x_{k+1} + c_{1,k+2}x_{k+2} + \cdots + c_{1n}x_n = d_1 \\
 x_2 + c_{2,k+1}x_{k+1} + c_{2,k+2}x_{k+2} + \cdots + c_{2n}x_n = d_2 \\
\\
 \cdot \cdot \cdot \cdot \cdot \cdot \\
\\
x_k + c_{k,k+1}x_{k+1} + c_{k,k+2}x_{k+2} + \cdots + c_{kn}x_n = d_k \\
0x_{k+1} + 0x_{k+2} + \cdots + 0x_n = d_{k+1} \\
0x_{k+1} + 0x_{k+2} + \cdots + 0x_n = d_{k+2} \\
\\
 \cdot \cdot \cdot \cdot \cdot \cdot \\
\\
0x_{k+1} + 0x_{k+2} + \cdots + 0x_n = d_m
\end{cases}$$

and the corresponding augmented matrix. If any one of the numbers $d_{k+1}, \ldots, d_m$ is different from zero, we obtain a contradiction and the above system is inconsistent. Hence, the equivalent original system is inconsistent. But if $d_{k+1} = d_{k+2} = \ldots = d_m = 0$, the last $m - k$ equations are satisfied and so are the first k equations, since we can solve them for $x_1, \ldots, x_k$ in terms of $x_{k+1}, \ldots, x_n$. Therefore the system is consistent if and only if

$$d_{k+1} = d_{k+2} = \cdots = d_m = 0.$$

Since $|I_k| \neq 0$, the rank of the coefficient matrix is equal to k, and so is the rank of the augmented matrix when $d_{k+1} = d_{k+2} = \cdots = d_m = 0$. But if any one of these $m - k$ numbers is different from zero, the rank of the augmented matrix is greater than k. Hence the system is consistent if and only if the ranks of the coefficient and augmented matrices are the same.

A.9 CONDITION FOR UNIQUENESS OF SOLUTION OF A LINEAR SYSTEM

Theorem A.7 [Theorem 4.2, page 73]. *A consistent system of linear equations $AX = B$ in n unknowns has a unique solution if and only if $r = n$.*

Proof. In the proof of Theorem A.6 if $AX = B$ is consistent and the rank equals k, we can solve for certain x_j in terms of d_j and $n - k$ remaining unknowns, that is,

$$x_j = d_j - c_{j,k+1}x_{k+1} - c_{j,k+2}x_{k+2} - \cdots - c_{jn}x_n, \qquad (j = 1, 2, \ldots, k).$$

If the system has a unique solution, then $n - k = 0$, and hence $k = n$ which means that the rank of the system equals the number of unknowns. Conversely, if the rank of the system equals the number of unknowns, then $k = n$ and hence $x_j = d_j$, which is a unique solution where $j = 1, 2, \cdots, n$.

A.10 RANK OF EQUIVALENT MATRICES

Theorem A.8 [Theorem 5.3, page 92]. *Equivalent matrices have the same rank.*

Outline of Proof.   The proof is split into two parts.

FIRST: Show that elementary operations (1) and (2) do not change a nonzero minor to zero, nor change a zero minor to one that is nonzero, and therefore does not alter the rank of a matrix.

SECOND: Show that elementary operation (3) does not increase the rank of a matrix A. If the rank r of A is as large as the order of A permits, then r cannot be increased. If the rank r of A is not as large as the order permits, then consider all submatrices of order $r + 1$ after elementary operation (3) has been applied. By a generalization of Exercise 16 of Section 3.4 and Theorem 3.13 the determinants of these submatrices can be expressed as $|S| + k|R|$, where S is a submatrix of A of order $r + 1$. Either R is a submatrix of A of order $r + 1$ with perhaps one line out of place or with two identical lines. In any event $|R| = 0$; also $|S| = 0$ because the rank of A was r. Therefore all submatrices of order $r + 1$ of the transformed matrix have a rank no greater than r, and thus the rank of A is no greater than r. Neither is the rank less than r because the inverse transformation (of the same type) would cause an increase in rank, which we have just shown is impossible.

A.11 DETERMINATION OF A SET OF PARAMETERS

Theorem A.9 [Theorem 6.1, page 106]. *A consistent system of linear equations of rank r can be solved for   r   unknowns, say* $x_{i_1}, x_{i_2}, \ldots, x_{i_r}$ *in terms of the remaining n − r unknowns, if and only if the submatrix of coefficients of* $x_{i_1}, x_{i_2}, \cdots, x_{i_r}$ *has rank   r.*

Outline of Proof. Assume that the rank of the submatrix of coefficients of r of the unknowns is r. We have already shown in the process of proving Theorem A.7 that these unknowns can be found in terms of the remaining $n - r$ unknowns.

Conversely, if we assume that r of the unknowns, $x_{i_1}, x_{i_2}, \cdots, x_{i_r}$, can be found in terms of the other $n - r$ unknowns, the rank of this system must be r. Hence, the rank of the coefficient matrix of this system is r, and by elementary operations, r can be shown to equal the rank of the submatrix of coefficients of $(x_{i_1}, x_{i_2}, \cdots, x_{i_r})$ in the original system.

A.12 CHARACTERISTIC VALUES OF REAL SYMMETRIC MATRICES

Theorem A.10 [Theorem 8.2, page 130]. *The characteristic values of real symmetric matrices are real numbers.*

Proof. The complex conjugate of $AX = \lambda X$ is $\bar{A}\bar{X} = \bar{\lambda}\bar{X}$ or $A\bar{X} = \bar{\lambda}\bar{X}$, since A is real. Premultiplying the first equation by $\bar{X}^{\mathbf{T}}$ and the last equation by $X^{\mathbf{T}}$ yields

$$\bar{X}^{\mathbf{T}}AX = \lambda\bar{X}^{\mathbf{T}}X \qquad \text{and} \qquad X^{\mathbf{T}}A\bar{X} = \bar{\lambda}X^{\mathbf{T}}\bar{X}.$$

Subtraction yields

$$\bar{X}^{\mathbf{T}}AX - X^{\mathbf{T}}A\bar{X} = \lambda\bar{X}^{\mathbf{T}}X - \bar{\lambda}X^{\mathbf{T}}\bar{X}.$$

But the left-hand side is zero because

$$\bar{X}^{\mathbf{T}}AX = (\bar{X}^{\mathbf{T}}AX)^{\mathbf{T}} = X^{\mathbf{T}}A^{\mathbf{T}}\bar{X} = X^{\mathbf{T}}A\bar{X}.$$

The first step is valid because $\bar{X}^{\mathbf{T}}AX$ is a 1 by 1 matrix, and in last step, we used the fact that A is symmetric. Therefore,

$$0 = \lambda\bar{X}^{\mathbf{T}}X - \bar{\lambda}X^{\mathbf{T}}\bar{X}.$$

But $\bar{X}^T X = X^T \bar{X}$ which is real and positive (that is, $(a + bi)(a - bi) = a^2 + b^2$); hence,

$$0 = (\lambda - \bar{\lambda})X^T\bar{X},$$

which implies $\lambda - \bar{\lambda} = 0$ and $\lambda = \bar{\lambda}$. Therefore λ must be real.

A.13 ORTHOGONAL CHARACTERISTIC VECTORS

Theorem A.11 [Theorem 8.3, page 130] *Any two characteristic vectors X_i, X_j, corresponding to two distinct characteristic values λ_i, λ_j of a real symmetric matrix are orthogonal.*

Proof. $AX_i = \lambda_i X_i$ and $AX_j = \lambda_j X_j$.
Hence,

$$X_j{}^T(AX_i) = X_j{}^T (\lambda_i X_i) \qquad \text{and} \qquad X_i{}^T (AX_j) = X_i{}^T (\lambda_j X_j).$$

Subtracting we get

$$X_j{}^T AX_i - X_i{}^T AX_j = \lambda_i X_j{}^T X_i - \lambda_j X_i{}^T X_j.$$

Since A is symmetric,

$$X_j{}^T AX_i = X_i{}^T AX_j;$$

hence, we have

$$0 = \lambda_i X_j{}^T X_i - \lambda_j X_i{}^T X_j.$$

But $X_j{}^T X_i = X_i{}^T X_j$, therefore

$$0 = (\lambda_i - \lambda_j)(X_j{}^T X_i),$$

and since $\lambda_i \neq \lambda_j$, then $X_j{}^T X_i = 0$.

A.14 A MATRIX OF ORTHOGONAL VECTORS

Theorem A.12 [Theorem 8.4, page 131]. *A matrix $T = [X_1 \mid X_2 \mid \ldots \mid X_n]$, where*

$$\{X_1, X_2, \cdots X_n\}$$

is an orthonormal set of n-dimensional, real vectors, is an orthogonal matrix. That is, $T^T = T^{-1}$.

Proof.

$$T^{\mathbf{T}}T = \begin{bmatrix} X_1^{\mathbf{T}} \\ \overline{}\overline{} \\ X_2^{\mathbf{T}} \\ \overline{}\overline{} \\ \cdot \\ \cdot \\ \cdot \\ \overline{}\overline{} \\ X_n^{\mathbf{T}} \end{bmatrix} [X_1 \mid X_2 \mid \cdots \mid X_n]$$

$$= \begin{bmatrix} X_1^{\mathbf{T}}X_1 & X_1^{\mathbf{T}}X_2 & \cdots & X_1^{\mathbf{T}}X_n \\ X_2^{\mathbf{T}}X_1 & X_2^{\mathbf{T}}X_2 & \cdots & X_2^{\mathbf{T}}X_n \\ \cdot & \cdot & \cdot & \cdot \\ \cdot & \cdot & \cdot & \cdot \\ \cdot & \cdot & \cdot & \cdot \\ X_n^{\mathbf{T}}X_1 & X_n^{\mathbf{T}}X_2 & \cdots & X_n^{\mathbf{T}}X_n \end{bmatrix} \quad \text{by matrix multiplication,}$$

$= I_n$ because $\{X_1, X_2, \cdots, X_n\}$ is an orthonormal set of real vectors.

By Exercise 12 of Section 5.1, $TT^{\mathbf{T}} = I_n$, and, therefore, since the inverse of a matrix is unique, $T^{\mathbf{T}} = T^{-1}$.

Answers to the Odd-Numbered Exercises

1.1 Page 5

1. (a)

x	y	$x \cup y$	$y \cap (x \cup y)$
0	0	0	0
0	1	1	1
1	0	1	0
1	1	1	1

Because there are a finite number of elements (two) we try all possibilities in the chart above and find that $y = y \cap (x \cup y)$ in all cases.

(b) ⸻ y ⸻ ⎡x⎤ ⸻ .
⎣y⎦

(c) "statement y" and ("statement x" or "statement y").

1.2 Page 10

1. 1 by 3; 2 by 4; 2 by 1; In the first matrix $3 = a_{13}$; the subscript 13 (one-three) is the address; In the second matrix $3 = a_{23}$, the subscript 23 is the address. In the third matrix $3 = a_{21}$, the subscript 21 is the address.

3. $\begin{bmatrix} i & 1 & 0 \\ 2 & 2 & \sqrt{3} \end{bmatrix}$.

5. $\begin{bmatrix} 0 & 0 & 0 & 0 & 0 \\ 0 & 0 & 0 & 0 & 0 \\ 0 & 0 & 0 & 0 & 0 \end{bmatrix}$.

7. (a) $x = 0$.

(b) t is arbitrary.

(c) Impossible (the orders are not the same).

163

(d) $x \geq 4$.

(e) Impossible (imaginary numbers are not ordered).

(f) Impossible ($4 \not< 4$).

9.

	#1	#2	#3	#4
#1	0	0	0	1
#2	1	0	1	1
#3	0	1	0	0
#4	1	0	0	0

11. $\begin{cases} x + 3y = 1, \\ 4x + 2y = 5, \\ 2x + 6y = 4. \end{cases}$

1.3 Page 18

1. (a) $(4, -4, 3, 4 + x)$;   (b) $(6, -9, 0, 12)$;

(c) Yes, if $x \geq 4$ because then each of the components of β are greater than or equal to the corresponding component of α;

(d) $(-2, -3, -9, 8 - 3x)$;   (e) $(10, -9, 9, 8 + 3x)$;

(f) $(2, -3, 0, 4) = \alpha$.

3. (a) Sum is $(-1, 2, 5)$;   dot product is 4;

(b) Sum is $(9, 1, 4, 1)$;   dot product is 16;

(c) Not possible because the orders are not the same.

5. $\dfrac{-1}{5\sqrt{2}}$;   the angle is obtuse.

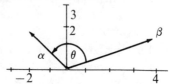

7. Proof:

STATEMENT	REASON
(1) $\alpha \cdot \beta = \lvert\alpha\rvert\ \lvert\beta\rvert \cos\theta$.	(1) Theorem 1.1, page 14.
(2) $0 = \lvert\alpha\rvert\ \lvert\beta\rvert \cos\theta$.	(2) Given that $\alpha \cdot \beta = 0$.
(3) Hence $\cos\theta = 0$.	(3) Given that $\lvert\alpha\rvert \neq 0$ and $\lvert\beta\rvert \neq 0$.
(4) Therefore α is perpendicular to β.	(4) Since $0° \leq \theta \leq 180°$, $\cos\theta = 0 \Rightarrow \theta = 90°$.

9. 30° east of north;   speed $= 25(1 + \sqrt{3})$.

11. 6 units.

13. $43 profit for the refining industry.

$20 profit for the utility industry.

2.1 Page 22

1. (*a*) Yes;   (*b*) No.

3. No.

2.2 Page 25

1. (*a*) $\begin{bmatrix} 4 & 4 \\ 3 & 5 \end{bmatrix}$;   (*b*) Impossible;   (*c*) $\begin{bmatrix} 3 & 4 \\ 1+\sqrt{2} & 0 \\ 6 & i \end{bmatrix}$;   (*d*) $\begin{bmatrix} 5 \\ 5 \\ 10 \end{bmatrix}$.

3. Proof:

	STATEMENT		REASON
(1)	$A + B = [a_{ij}]_{(m,n)} + [b_{ij}]_{(m,n)}$	(1)	Change in notation.
(2)	$= [a_{ij} + b_{ij}]_{(m,n)}$	(2)	Definition 2.2.
(3)	$= [b_{ij} + a_{ij}]_{(m,n)}$	(3)	Commutative property for addition of scalars.
(4)	$= [b_{ij}]_{(m,n)} + [a_{ij}]_{(m,n)}$	(4)	Definition 2.2.
(5)	$= B + A$	(5)	Change in notation.

2.3 Page 27

1. (*a*) $\begin{bmatrix} 6 & -3 \\ -9 & -12 \end{bmatrix}$;   (*b*) $\begin{bmatrix} 4 & 0 \\ 2 & -6 \end{bmatrix}$;   (*c*) $\begin{bmatrix} -2 & 1 \\ 3 & 4 \end{bmatrix}$;

(*d*) $\begin{bmatrix} -4 & -1 \\ -6 & 5 \end{bmatrix}$;   (*e*) $\begin{bmatrix} -5 & 2 \\ \frac{11}{2} & \frac{19}{2} \end{bmatrix}$;   (*f*) $\begin{bmatrix} 4 & -1 \\ -2 & -7 \end{bmatrix}$;

(*g*) $\begin{bmatrix} 3 & -\frac{1}{2} \\ -\frac{1}{2} & -5 \end{bmatrix}$.

3. No.

5. λI; $\begin{bmatrix} \lambda & 0 & 0 & 0 \\ 0 & \lambda & 0 & 0 \\ 0 & 0 & \lambda & 0 \\ 0 & 0 & 0 & \lambda \end{bmatrix}$.

7. Proof:

STATEMENT		REASON
(1) $Ac = [a_{ij}c]_{(m,n)}$	(1)	Definition 2.3.
(2) $\quad = [ca_{ij}]_{(m,n)}$	(2)	Multiplication of scalars is commutative.
(3) $\quad = cA$	(3)	Definition 2.3.

2.4 Page 35

1. (a) $\begin{bmatrix} 2 & 1 \\ 8 & -1 \end{bmatrix}$; (b) $\begin{bmatrix} 2 \\ -6 \end{bmatrix}$; (c)[8 2]; (d) $\begin{bmatrix} 4 & 2 \\ 3 & 1 \end{bmatrix}$;

   (e) Impossible, matrices are not conformable; (f) $\begin{bmatrix} 6 & -2 \\ 9 & -3 \end{bmatrix}$.

3. (a) $n = r$; (b) m by t; (c) $t = m$; (d) r by n;
   (e) $m = n = r = t$.

5. $m = p$.

7. $AX = B$, where $A = \begin{bmatrix} 1 & 1 & 1 \\ 1 & -1 & 2 \\ 2 & 0 & 1 \end{bmatrix}$, $X = \begin{bmatrix} x_1 \\ x_2 \\ x_3 \end{bmatrix}$, $B = \begin{bmatrix} 4 \\ 9 \\ 6 \end{bmatrix}$.

9. $AN = \begin{bmatrix} 16 \\ 38 \end{bmatrix}$. The entries of this matrix represent the total number of Gadget R and Gadget S that can be produced in one week by the two factories.

11. (a) $A + A^2 = \begin{bmatrix} 0 & 1 & 1 & 2 \\ 1 & 0 & 2 & 2 \\ 0 & 1 & 0 & 1 \\ 1 & 1 & 1 & 0 \end{bmatrix}$; (b) $\begin{cases} \text{\#2 has five influence channels,} \\ \text{\#1 has four influence channels,} \\ \text{\#4 has three influence channels,} \\ \text{\#3 has two influence channels.} \end{cases}$

13.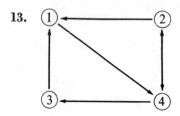

2.5 Page 44

1. (a) The union is $\begin{bmatrix} 0 & 1 \\ 1 & 1 \end{bmatrix}$; the intersection is $\begin{bmatrix} 0 & 0 \\ 1 & 0 \end{bmatrix}$;

   (b) $\begin{bmatrix} 0 & 0 \\ 0 & 0 \end{bmatrix}$ for $\cup$; $\begin{bmatrix} 1 & 1 \\ 1 & 1 \end{bmatrix}$ for $\cap$;

(c) Proof: STATEMENT   REASON

(1) $A \cup B = [a_{ij} \cup b_{ij}]_{(2,2)}$ (1) Definition 2.7.

(2) $= [b_{ij} \cup a_{ij}]_{(2,2)}$ (2) Definition of Example 1 of Section 1.1.

(3) $= B \cup A.$ (3) Definition 2.7.

3. Proof of commutativity: By Definition 2.8 and the commutative law for matrix addition,

$$A \textcircled{j} B = \frac{AB + BA}{2} = \frac{BA + AB}{2} = B \textcircled{j} A.$$

The Jordan product is not associative because

$$(A \textcircled{j} B) \textcircled{j} C = \tfrac{1}{4}(ABC + BAC + CAB + CBA),$$

whereas

$$A \textcircled{j} (B \textcircled{j} C) = \tfrac{1}{4}(ABC + ACB + BCA + CBA).$$

5. (a) $f(x, y) = \{x' \cap (x \cup y)\} \cup \{(y \cup x) \cap (x' \cup y)\}.$
   (b) $f(x, y) = y.$

9. Statements:
   (a) Starting with the right-hand side:
   $B \times (A \times C) + C \times (B \times A)$
   $= B(A \times C) - (A \times C)B + C(B \times A) - (B \times A)C$
   $= B(AC - CA) - (AC - CA)B + C(BA - AB)$
   $\quad - (BA - AB)C$
   $= BAC - BCA - ACB + CAB + CBA - CAB$
   $\quad - BAC + ABC$
   $= -BCA - ACB + CBA + ABC$
   $= ABC - ACB - BCA + CBA$
   $= A(BC - CB) - (BC - CB)A$
   $= A(B \times C) - (B \times C)A$
   $= A \times (B \times C).$
   (b) $A \times (B + C) = A(B + C) - (B + C)A$
   $\qquad\qquad\quad\ = AB + AC - BA - CA$
   $\qquad\qquad\quad\ = AB - BA + AC - CA$
   $\qquad\qquad\quad\ = (A \times B) + (A \times C).$

(c) $(A + B) \times C = -\left(C \times (A + B)\right)$

$$= -\left((C \times A) + (C \times B)\right)$$
$$= -(C \times A) - (C \times B)$$
$$= (A \times C) + (B \times C).$$

(d) $c(A \times B) = c(AB - BA)$
$$= c(AB) - c(BA)$$
$$= (cA)B - (cB)A$$
$$= (cA)B - (Bc)A$$
$$= (cA)B - B(cA)$$
$$= cA \times B.$$

Similarly, it can be shown that $c(A \times B) = A \times (cB)$.

3.2 Page 52

1. (a) symmetric;   (b) neither; not a square matrix;
   (c) skew-symmetric;   (d) neither; $A \neq A^T$; $A \neq -\bar{A}^T$.

3. (b) $\begin{bmatrix} 2 & 0 \\ 7 & 4 \\ 6 & 8 \end{bmatrix} = (AB)^T = B^T A^T$.

5. Statements:

$$(A - B)^T = (A + (-B))^T = A^T + (-B)^T = A^T - B^T.$$

7. Matrices in parts (c) and (d) are Hermitian. Matrix in part (e) is skew-Hermitian. Matrices in parts (a) and (b) are neither; matrix in part (a) is not a square matrix and in part (b) $A \neq \bar{A}^T$ and $A \neq -\bar{A}^T$.

9. Statements: $(A + A^T)^T = A^T + (A^T)^T = A^T + A = A + A^T$.

11. Statements: $(A^2)^T = (AA)^T = A^T A^T = (-A)(-A)$
$$= \{A(-1)\}(-A) = A\{(-1)(-A)\} = AA = A^2.$$

13. Statements: $AA^T = -A^T A^T = \{A^T(-1)\}A^T = A^T\{-A^T\} = A^T A$.

15. Statements:
   (a) $(A^T)^T = ([a_{ij}]^T_{(m,n)})^T = ([a_{ji}]_{(n,m)})^T = [a_{ij}]_{(m,n)} = A$;
   (b) $(A + B)^T = [a_{ij} + b_{ij}]^T_{(m,n)} = [a_{ji} + b_{ji}]_{(n,m)} = A^T + B^T$;
   (c) $(cA)^T = [ca_{ij}]^T_{(m,n)} = [ca_{ji}]_{(n,m)} = c[a_{ji}]_{(n,m)} = cA^T$.

17. $f(x, y, z) = \begin{bmatrix} x & y & z \end{bmatrix} \begin{bmatrix} 1 & -2 & 0 \\ -2 & 1 & 3 \\ 0 & 3 & 1 \end{bmatrix} \begin{bmatrix} x \\ y \\ z \end{bmatrix}$.

3.3 Page 56

1. 6.

3. Statements:

(a) tr $A = a_{11} + a_{22} + \cdots + a_{nn} = $ tr $A^\mathbf{T}$;

(b) tr $(A + B) = (a_{11} + b_{11}) + (a_{22} + b_{22}) + \cdots + (a_{nn} + b_{nn})$
$$= a_{11} + a_{22} + \cdots + a_{nn} + b_{11} + b_{22} + \cdots + b_{nn}$$
$$= \text{tr } A + \text{tr } B;$$

(c) tr $(cA) = ca_{11} + ca_{22} + \cdots + ca_{nn}$
$$= c(a_{11} + a_{22} + \cdots + a_{nn}) = c \text{ tr } A.$$

5. tr $(A + A^2 + A^3) = 42$ represents the total possible round trips originating from all cities and passing through 1 *or* 2 of the other cities.

3.4 Page 64

1. (a) $|A| = 3\begin{vmatrix} 9 & -2 \\ 2 & 1 \end{vmatrix} - 6\begin{vmatrix} 1 & -4 \\ 2 & 1 \end{vmatrix} + (-1)\begin{vmatrix} 1 & -4 \\ 9 & -2 \end{vmatrix} = -49;$

(b) -49; (c) -49; (d) $(-1)^5 \begin{vmatrix} 3 & -4 \\ 6 & -2 \end{vmatrix} = -18;$

(e) $\begin{vmatrix} 6 & -2 \\ -1 & 1 \end{vmatrix} = 4.$

3. 15.

5. $+4$;   cofactor of $a_{ij} = (-1)^{i+j}$(minor of a_{ij}).

7. (a) $(a_{11} \; a_{22} \; a_{33})$;   (b) $(a_{11} \; a_{22} \; \cdots \; a_{nn})$.

9. -8 by Theorem 3.10.

11. 108;   det $(3A) = 3^3$ det $A = 108$ (by application of Theorem 3.13 three times).

13. (a) Proof: Let B be the matrix created by multiplying every entry of the ith row of A by a scalar c.

$$\det B = \det \begin{bmatrix} a_{11} & \cdots & a_{1n} \\ \cdot & \cdot & \cdot \\ \cdot & \cdot & \cdot \\ \cdot & \cdot & \cdot \\ ca_{i1} & \cdots & ca_{in} \\ \cdot & \cdot & \cdot \\ \cdot & \cdot & \cdot \\ \cdot & \cdot & \cdot \\ a_{n1} & \cdots & a_{nn} \end{bmatrix}$$

$$= ca_{i1}A_{i1} + ca_{i2}A_{i2} + \cdots + ca_{in}A_{in} \text{ by Definition 3.10.}$$

If c is factored out, there results $c \det A$ by Definition 3.10.

(b) Proof:

STATEMENT	REASON

$$\det B = \det \begin{bmatrix} a_{11} & \cdots & a_{1n} \\ \cdot & \cdot & \cdot \\ \cdot & \cdot & \cdot \\ \cdot & \cdot & \cdot \\ a_{i1} & \cdots & a_{in} \\ \cdot & \cdot & \cdot \\ \cdot & \cdot & \cdot \\ ca_{i1} + a_{k1} & \cdots & ca_{in} + a_{kn} \\ \cdot & \cdot & \cdot \\ \cdot & \cdot & \cdot \\ \cdot & \cdot & \cdot \\ a_{n1} & \cdots & a_{nn} \end{bmatrix}$$ Given.

$= (ca_{i1} + a_{k1})A_{k1} + \cdots$
$\quad + (ca_{in} + a_{kn})A_{kn}$ Definition 3.10.

$= (ca_{i1}A_{k1} + \cdots + ca_{in}A_{kn})$
$\quad + (a_{k1}A_{k1} + \cdots + a_{kn}A_{kn})$ Rearranging terms.

$= c(a_{i1}A_{k1} + \cdots + a_{in}A_{kn})$
$\quad + \det A$ Definition 3.10.

$= 0 + \det A$ Theorem 3.12.

15. $\begin{bmatrix} 1 & 1 & 0 \\ 1 & 3 & 2 \\ 0 & 1 & 1 \end{bmatrix}$ is one counterexample.

17. Outline of proof: Rearrange terms and factor appropriately to obtain an expression equivalent to that found in Definition 3.10; -11.

4.1   Page 77

1. (a) 2;   (b) 3;   (c) 1;   (d) 0;   (e) 1.

3. n.

5. (a) Rank of A is less than 4.
   (b) Rank of A is 4.

7. (d) and (e).

9. Rank of augmented matrix is 5. There must be at least 5 equations.

11. r is the order of a submatrix of A and the order of A is m by n, hence $r \le n$ and $r \le m$.

5.1 Page 89

1. $\text{adj } A = \begin{bmatrix} 1 & -3 \\ 4 & 2 \end{bmatrix}$; $\text{adj } B = \begin{bmatrix} -1 & 3 & 0 \\ 2 & -1 & 0 \\ 2 & -1 & -5 \end{bmatrix}$;

$\text{adj } C = \begin{bmatrix} -4 & 0 & 0 & 2 \\ 0 & 0 & -2 & 0 \\ 0 & -4 & 0 & 0 \\ 0 & 0 & 0 & -2 \end{bmatrix}.$

3. $A^{-1} = \begin{bmatrix} \frac{3}{10} & \frac{1}{10} \\ -\frac{4}{10} & \frac{2}{10} \end{bmatrix}$; B^{-1} does not exist because B is singular.

$C^{-1} = \frac{1}{7} \begin{bmatrix} 2 & -3 & 0 \\ -1 & -2 & 7 \\ 1 & 2 & 0 \end{bmatrix}$; $D^{-1} = \frac{1}{19} \begin{bmatrix} 4 & 4 & -1 \\ -1 & -1 & 5 \\ 9 & -10 & -7 \end{bmatrix}.$

5. They are equal.

7. $A^{-2} = \frac{1}{9} \begin{bmatrix} 1 & -8 \\ 0 & 9 \end{bmatrix}$; $A^{-3} = \frac{1}{27} \begin{bmatrix} 1 & -26 \\ 0 & 27 \end{bmatrix}.$

9. Proof:

STATEMENT	REASON
(1) $AA^{-1} = A^{-1}A = I$ $\Rightarrow A$ is the inverse of A^{-1}.	(1) Definition 5.1.
(2) $(A^{-1})^{-1}A^{-1} = (A^{-1})^{-1}A^{-1} = I$ $\Rightarrow (A^{-1})^{-1}$ is the inverse of A^{-1}.	(2) Definition 5.1.
(3) $A = (A^{-1})^{-1}.$	(3) Theorem 5.2.

11. Statements: $(AB)(B^{-1}A^{-1}) = A\{(BB^{-1})A^{-1}\} = A\{IA^{-1}\} = AA^{-1}$ $= I$. Likewise $(B^{-1}A^{-1})(AB) = I$. Therefore the inverse of AB is $B^{-1}A^{-1}$.

13. Statements: A is nonsingular $\Rightarrow A^{-1}$ exists. Therefore, $BA = CA$ $\Rightarrow (BA)A^{-1} = (CA)A^{-1} \Rightarrow B(AA^{-1}) = C(AA^{-1}) \Rightarrow BI = CI \Rightarrow$ $B = C$.

15. $X = \begin{bmatrix} 2 \\ 10 \end{bmatrix}.$

5.2 Page 96

1. $x = -1, y = 2$.

3. No;   the rank is 2.

5. (a) 2; (b) 2; (c) 2.

7. (a) $E = \begin{bmatrix} 0 & 1 \\ 1 & 0 \end{bmatrix}$; (b) $E = \begin{bmatrix} 1 & 0 & 0 \\ 0 & 1 & 0 \\ 0 & 0 & 7 \end{bmatrix}$; (c) $E = \begin{bmatrix} 1 & 0 \\ 4 & 1 \end{bmatrix}$;

(d) $E = \begin{bmatrix} 1 & 0 & 0 \\ 0 & 1 & 0 \\ 5 & 0 & 1 \end{bmatrix}$.

9. (a) $\frac{1}{15} \begin{bmatrix} 2 & 1 \\ -7 & 4 \end{bmatrix}$; (b) $\begin{bmatrix} 0 & 0 & 1 \\ 0 & 1 & -3 \\ 1 & 0 & -1 \end{bmatrix}$; (c) $\frac{1}{4} \begin{bmatrix} 3 & -3 & -4 \\ 0 & 4 & 0 \\ -2 & 2 & 4 \end{bmatrix}$.

11. $\frac{25}{4}$ tons of P;   5 tons of Q.

13. $\frac{14}{13}$ barrels of P; $\frac{16}{13}$ barrels of Q; $\frac{8}{13}$ barrels of R.

6.1 Page 104

1. $\begin{cases} x_1 = 3, \\ x_2 = 1 + x_3; \end{cases}$ $(3, 2, 1)$.

3. $\begin{cases} x_1 = 0 + x_2, \\ x_3 = 4 - 2x_2; \end{cases}$ $(1, 1, 2)$.

5. $\begin{cases} x_1 = -3 + 3x_3 - 3x_4, \\ x_2 = -3 + 2x_3 - 2x_4; \end{cases}$ $(-3, -3, 0, 0)$.

7. $x_2 = 4 - 2x_1$; $(1, 2)$.

9. $\begin{cases} x_1 = \frac{2}{5}x_3, \\ x_2 = \frac{4}{5}x_3; \end{cases}$ $(\frac{2}{5}, \frac{4}{5}, 1)$.

11. The rank of the system is less than the number of unknowns.

13. (a) There is no unique answer; 4 type 1 trucks must be sent, and if k type 3 trucks are sent, then $(8 - k)$ type 2 trucks must be sent.

(b) There will be exactly 9 solutions because in order to send fully loaded trucks, k must be an integer such that $0 \le k \le 8$.

6.2   Page 110

1. No; Theorem 6.1.

3. (a) Let $x_1 = x_2 = 0$. A basic solution would then be $(0, 0, 4, 5)$.
(b) Any two unknowns can be found in terms of any other two except x_1 and x_2 cannot be found in terms of x_3 and x_4.

5. (a) $\begin{cases} z_1 + 2z_2 + z_3 \quad\quad = 5, \\ z_1 + 3z_2 \quad\quad + z_4 = 7, \end{cases}$ where $\begin{cases} z_3 \geq 0, \\ z_4 \geq 0, \end{cases}$

(b) $\begin{cases} z_1 + z_2 + z_3 \quad\quad\quad = 4, \\ z_1 + 4z_2 \quad\quad + z_4 \quad\quad = 7, \\ 2z_2 \quad\quad\quad + z_5 = 3, \end{cases}$ where $\begin{cases} z_3 \geq 0, \\ z_4 \geq 0, \\ z_5 \geq 0. \end{cases}$

7. Program A twice. Program B four times.

9. 60 lbs of brand P and none of brand Q.

7.1   Page 118

1. (a) $\begin{bmatrix} -2 \\ 4 \end{bmatrix}$;   (b) $\begin{bmatrix} 0 \\ -6 \end{bmatrix}$;   (c) $\begin{bmatrix} 4 \\ 4 \end{bmatrix}$.

3. Domain: The set of real numbers.
Range: The set of 2-dimensional vectors with real components and with a second component which is twice the first.

5. Domain: The set of 2-dimensional vectors with real components.
Range: The set of vectors with zero as the first component and any real number as the second component.

7. Domain: The set of real numbers.
Range: The set of pure imaginary numbers bi where $b > 0$ and the set of real numbers between 0 and 1 inclusive.

9. Domain: The set of real numbers.
Range: The set of complex numbers $1 + bi$ where $0 < b \leq 1$ and the set of real numbers greater than or equal to one.

11. A stretching (twice the magnitude) of a reflection through the origin.

13. A shear parallel to the x_1-axis.

15. A stretching (twice the magnitude) of a rotation of 90°.

7.2   Page 122

1. $A = \begin{bmatrix} k & 0 \\ 0 & k \end{bmatrix}$.

3. $(a)\ A = \begin{bmatrix} 1 & 0 \\ 0 & 0 \end{bmatrix};$ $(b)\ A = \begin{bmatrix} 2 & 0 \\ 0 & -2 \end{bmatrix};$ $(c)\ A = \begin{bmatrix} \dfrac{\sqrt{2}}{2} & \dfrac{-\sqrt{2}}{2} \\ \dfrac{\sqrt{2}}{2} & \dfrac{\sqrt{2}}{2} \end{bmatrix}.$

7.3   Page 124

1. $T\{(a_1, b_1) + (a_2, b_2)\} = (a_1 + a_2, 2)$, whereas
$T(a_1, b_1) + T(a_2, b_2) = (a_1 + a_2, 4)$,
and they are not equal.
Also, $T(ka_1, kb_1) = (ka_1, 2)$, whereas
$k\{T(a_1, b_1)\} = (ka_1, 2k)$,
and they are not equal.

3. $T\{(a_1, b_1) + (a_2, b_2)\} = T(a_1, b_1) + T(a_2, b_2)$
$\qquad\qquad\qquad\qquad = (-2[a_1 + a_2], 2[b_1 + b_2])$.
Also, $Tk(a, b) = kT(a, b)$
$\qquad\qquad = (-2ka, 2kb)$.

5. Theorem 2.5 justifies $A(\alpha + \beta) = A\alpha + A\beta$; then the fourth, first, and third parts of Theorem 2.6 justify $A(k\alpha) = k(A\alpha)$.

8.1   Page 136

1. $(a)\ \lambda^2 - 7\lambda + 6 = 0;$ $\lambda_1 = 1,$ $\lambda_2 = 6;$
$\quad X_1 = \pm \dfrac{1}{\sqrt{5}} \begin{bmatrix} 2 \\ 1 \end{bmatrix},$ $X_2 = \pm \dfrac{1}{\sqrt{5}} \begin{bmatrix} 1 \\ -2 \end{bmatrix};$

$(b)\ \lambda^2 - 6\lambda + 8 = 0;$ $\lambda_1 = 2, \lambda_2 = 4;$
$\quad X_1 = \pm \dfrac{1}{\sqrt{10}} \begin{bmatrix} 3 \\ 1 \end{bmatrix},$ $X_2 = \pm \dfrac{1}{\sqrt{2}} \begin{bmatrix} 1 \\ 1 \end{bmatrix};$

$(c)\ \lambda^2 - 6\lambda + 8 = 0;$ $\lambda_1 = 2, \lambda_2 = 4;$
$\quad X_1 = \pm \dfrac{1}{\sqrt{2}} \begin{bmatrix} -1 \\ 1 \end{bmatrix},$ $X_2 = \pm \dfrac{1}{\sqrt{10}} \begin{bmatrix} -3 \\ 1 \end{bmatrix};$

$(d)\ (\lambda - 3)(\lambda - 1)\lambda = 0;$ $\lambda_1 = 3, \lambda_2 = 1, \lambda_3 = 0;$
$\quad X_1 = \pm \dfrac{1}{\sqrt{6}} \begin{bmatrix} +2 \\ 1 \\ 1 \end{bmatrix},$ $X_2 = \pm \dfrac{1}{\sqrt{2}} \begin{bmatrix} 0 \\ -1 \\ 1 \end{bmatrix},$

$\qquad\qquad\qquad\qquad\qquad X_3 = \pm \dfrac{1}{\sqrt{3}} \begin{bmatrix} -1 \\ 1 \\ 1 \end{bmatrix};$

(e) $\lambda^2 - 6\lambda + 5 = 0$; $\lambda_1 = 5, \lambda_2 = 1$;

$$X_1 = \pm \frac{1}{\sqrt{10}} \begin{bmatrix} 3 \\ 1 \end{bmatrix}, \quad X_2 = \pm \frac{1}{\sqrt{2}} \begin{bmatrix} -1 \\ 1 \end{bmatrix};$$

(f) $\lambda^2 - 4 = 0$; $\lambda_1 = 2, \lambda_2 = -2$;

$$X_1 = \pm \frac{1}{\sqrt{2}} \begin{bmatrix} 1 \\ 1 \end{bmatrix}, \quad X_2 = \pm \frac{1}{\sqrt{2}} \begin{bmatrix} -1 \\ 1 \end{bmatrix}.$$

3. $\begin{bmatrix} \frac{3}{2} & \frac{1}{2} \\ \frac{1}{2} & \frac{3}{2} \end{bmatrix}$, $\begin{bmatrix} \frac{3}{2} & \frac{1}{4} \\ 1 & \frac{3}{2} \end{bmatrix}$, and $\begin{bmatrix} 1 & 0 \\ 0 & 2 \end{bmatrix}$ are three examples; there are others.

5. n characteristic values.

7. (a) $\begin{bmatrix} 3 & 5 \\ -1 & 2 \end{bmatrix}$ is one example.   (b) $\begin{bmatrix} 3 & i \\ i & 2 \end{bmatrix}$ is one example.

9. Not possible for parts (b) (c) and (e) because the matrices are not symmetric.

11. Exercise 1(b) of this section is one counterexample. This conjecture does not require that the matrix be symmetric as does Theorem 8.3.

13. The matrices of (a), (c), and (d) are orthogonal, that of (b) is not.

15. $\begin{bmatrix} k_1 \\ k_1 \end{bmatrix}$ or $\begin{bmatrix} 2k_2 \\ k_2 \end{bmatrix}$ where k_1 and k_2 are arbitrary.

17. Let a, a', b, and b' be expressed as column vectors. Show that $Ta = a'$ and $Tb = b'$; $Oc = (2, 0)$; $Og = (\frac{9}{5}, 6)$.

8.2   Page 147

1. The following answers are not unique.
   (a) $6u^2 + v^2 = 6$,   (b) $3u^2 - 1v^2 = 1$,
   (c) $6u^2 - 7v^2 = 1$,   (d) $u^2 + 3v^2 = 12$,
   (e) $3u^2 + 5v^2 = 15$,   (f) $\frac{1}{2}u^2 - \frac{1}{2}v^2 = 4$.

3. (a) $\begin{bmatrix} 3 & 0 \\ 0 & -1 \end{bmatrix}$ or $\begin{bmatrix} -1 & 0 \\ 0 & 3 \end{bmatrix}$.

   (b) There is no matrix T such that $T^{-1}AT$ equals a diagonal matrix.

   (c) $\begin{bmatrix} 1 & 0 & 0 \\ 0 & 0 & 0 \\ 0 & 0 & 6 \end{bmatrix}$ is one of six answers.

5. (a) The system is stable because $\lambda_1 = \frac{2}{3}, \lambda_2 = \frac{1}{2}$, and both are less than one.

(b) Both characteristic values may be less than one; Both are not less than zero.

7. (a) The hypothesis of Theorem 8.5 requires the symmetry of A, and the conclusion of Theorem 8.5 is that A is congruent to a diagonal matrix.

(b) Statements of proof: Given: $P = [X_1 \mid X_2 \mid \cdots \mid X_n]$ where X_i are the characteristic vectors corresponding to the n distinct characteristic values λ_i of matrix A.

$$AP = A[X_1 \mid X_2 \mid \cdots \mid X_n] = [AX_1 \mid AX_2 \mid \cdots \mid AX_n]$$
$$= [\lambda_1 X_1 \mid \lambda_2 X_2 \mid \cdots \mid \lambda_n X_n]$$
$$= [X_1\lambda_1 \mid X_2\lambda_2 \mid \cdots \mid X_n\lambda_n]$$

$$= [X_1 \mid X_2 \mid \cdots \mid X_n] \begin{bmatrix} \lambda_1 & \cdots & & & 0 \\ & \cdot\, \lambda_2 & & & \cdot \\ & & \cdot & & \\ & & & \cdot & \\ & & & & \cdot \\ 0 & & \cdots & & \lambda_n \end{bmatrix}$$

$$= P \begin{bmatrix} \lambda_1 & \cdots & & & 0 \\ & \cdot\, \lambda_2 & & & \cdot \\ & & \cdot & & \\ & & & \cdot & \\ & & & & \cdot \\ 0 & & \cdots & & \lambda_n \end{bmatrix}.$$

Premultiplying by P^{-1} yields $P^{-1}AP = \begin{bmatrix} \lambda_1 & \cdots & & & 0 \\ & \cdot\, \lambda_2 & & & \cdot \\ & & \cdot & & \\ & & & \cdot & \cdot \\ & & & & \cdot \\ 0 & & \cdots & & \lambda_n \end{bmatrix}.$

P^{-1} exists because we were given that P is nonsingular.

9. Statements: $P^{-1}AP = B$ implies
$$|B - \lambda I| = |P^{-1}AP - \lambda I| = |P^{-1}AP - \lambda P^{-1}IP|$$
$$= |P^{-1}(A - \lambda I)P| = |P^{-1}|\,|A - \lambda I|\,|P|$$
$$= |A - \lambda I|\,|P|\,|P^{-1}| = |A - \lambda I|(1) = |A - \lambda I|.$$

11. $x_n = 2^n$.

13. (a) Let "$\overset{c}{=}$" represent "is congruent to."

(1) $I^{T}AI = A \Rightarrow A \overset{c}{=} A.$

(2) $P^{T}AP = B \Rightarrow A = (P^{T})^{-1}BP^{-1}$
$$\Rightarrow A = (P^{-1})^{T}B(P^{-1}) \therefore (A \overset{c}{=} B) \Rightarrow (B \overset{c}{=} A).$$

(3) $Q^{T}AQ = B$ and $P^{T}BP = C \Rightarrow P^{T}(Q^{T}AQ)P = C$
$$\Rightarrow (P^{T}Q^{T})A(QP) = C$$
$$\Rightarrow (QP)^{T}A(QP) = C$$
$$\therefore (A \overset{c}{=} B \text{ and } B \overset{c}{=} C) \Rightarrow (A \overset{c}{=} C).$$

(b) Proof is similar to proof of part (a).

Index